Internet Links
for Science Education

Student–Scientist Partnerships

INNOVATIONS IN SCIENCE EDUCATION AND TECHNOLOGY

Series Editor:

Karen C. Cohen, Harvard University, Cambridge, Massachusetts

The Hidden Curriculum—Faculty-Made Tests in Science
Part 1: Lower-Division Courses
Part 2: Upper-Division Courses
Sheila Tobias and Jacqueline Raphael

Internet Links for Science Education: Student–Scientist Partnerships
Edited by Karen C. Cohen

Web-Teaching: A Guide to Designing Interactive Teaching for the World Wide Web
David W. Brooks

A Continuation Order Plan is available for this series. A continuation order will bring delivery of each new volume immediately upon publication. Volumes are billed only upon actual shipment. For further information please contact the publisher.

Internet Links for Science Education

Student–Scientist Partnerships

Edited by

Karen C. Cohen

Harvard University
Graduate School of Education
Cambridge, Massachusetts

PLENUM PRESS • NEW YORK AND LONDON

Library of Congress Cataloging-in-Publication Data

Internet links for science education : student-scientist partnerships /
  edited by Karen C. Cohen.
        p.   cm. -- (Innovations in science education and technology)
  Includes bibliographical references and index.
  ISBN 0-306-45558-7
  1. Science--Study and teaching. 2. Educational innovations.
3. Computer network resources. 4. Internet (Computer network)
I. Cohen, Karen C. II. Series.
LB1585.I52 1997
372.3'5'044--dc21   97-5780
                                                                        CIP

Further information on student–scientist partnerships, can be found at the
TERC web site–http://www.terc.edu/ssp/ssp.html

ISBN 0-306-45558-7

© 1997 Plenum Press, New York
A Division of Plenum Publishing Corporation
233 Spring Street, New York, N.Y. 10013

http://www.plenum.com

10 9 8 7 6 5 4 3 2 1

Printed in the United States of America

Contributors

Ivar G. Babb National Undersea Research Center, University of Connecticut at Avery Point, Groton, CT 06340

Daniel Barstow, TERC, 2067 Massachusetts Avenue, Cambridge, MA 02140

Mimi L. Becker, Department of Natural Resources, 215 James Hall, University of New Hampshire, Durham, NH 03824

Tommie R. Blackwell, The Space Camp Program, U.S. Space and Rocket Center, Huntsville, AL 35807

Rick Bonney, Cornell Laboratory of Ornithology, 159 Sapsucker Woods Road, Ithaca, NY 14850

Thomas G. Carter, ADA Technologies, 304 Inverness Way, South, Suite 365, Englewood, CO 80112

Karen C. Cohen, Educational Technology Center, Harvard University, Graduate School of Education, Cambridge, MA 02138

Russell G. Congalton Department of Natural Resources, 215 James Hall, University of New Hampshire, Durham, NH 03824

André A. Dhondt, Cornell Laboratory of Ornithology, 159 Sapsucker Woods Road, Ithaca, NY 14850

Cheick Diarra, Jet Propulsion Laboratory, NASA, 4800 Oak Grove Drive, Pasadena, CA 91109

Daniel C. Edelson, Northwestern University, Evanston, IL 60204

Barry J. Fishman, Northwestern University, Evanston, IL 60204

Patrick D. French, ADA Technologies, 304 Inverness Way, South, Suite 365, Englewood, CO 80112

Louis M. Gomez, Northwestern University, Evanston, IL 60204

Douglas N. Gordin, Northwestern University, Evanston, IL 60204

Richard Greenberg, University of Arizona, Tucson, AZ 87519-0092

Anne Hardison, The GLOBE Program, 744 Jackson Place, Washington, DC 20503

Alice B. Kreuger, Eisenhower High Plains Consortium for Mathematics and Science at the Mid-continent Regional Education Laboratory, 2550 South Parker Road, Aurora, CO 80014

Danielle Miller, The GLOBE Program, 744 Jackson Place, Washington, DC 20503

Forrest M. Mims III, 433 Twin Oak Road, Seguin, TX 78155

M. Patricia Morse, Northeastern University, Marine Science Center, Nahant, MA 01908; *present address:* Division of Elementary, Secondary and Informal Education, National Science Foundation, 4201 Wilson Boulevard, Arlington, VA 22230

Rachel Nixon, EARTHWATCH, 680 Mt. Auburn Street, Box 9104, Watertown, MA 02272

D. Kevin O'Neill, Northwestern University, Evanston, IL 60204

Roy D. Pea, SRI International, 333 Ravenswood Avenue, Menlo Park, CA 94025

Barrett N. Rock, Complex Systems Research Center, University of New Hampshire, Durham, NH 03824

Nora H. Sabelli Division of Research, Evaluation, and Communication, National Science Foundation, 4201 Wilson Boulevard, Arlington, VA 22230

Peter M. Scheifele, National Undersea Research Center, University of Connecticut at Avery Point, Groton, CT 06340

Domenic Tedeschi, Norwich Free Academy, Norwich, CT 06360

Robert F. Tinker, The Concord Consortium, 37 Thoreau Street, Concord, MA 01742

Foreword

Science teachers come in many varieties, but they share a common goal: to nurture learners. Over the past decade, we have learned a great deal about how to do this effectively. Of all this new (and some not so new) knowledge, what strikes me as most important is that learning occurs best within a context. Still, as obvious as that may seem, it is relatively rare in our high school science classrooms. The problem, of course, is that it is not easy to create a learning experience with hands-on relevance to the science under discussion.

Science teachers, in addition to not having the the time, for the most part do not have the expertise or readily available resources. The solution lies in finding ways to bring scientists into the teaching/learning equation. Scientists teamed with teachers and their students represent a very real and rich opportunity to involve students in real science as practiced.

Imagine a research book that gives examples of honest, science-research experiences for science-oriented students. What's more, imagine a book that includes examples where students are collaborating with scientists from all over the world on research projects, in person or via the Internet. *Internet Links for Science Education* does just that. It explores the role of the Internet and technology in working student–scientist partnerships.

The various partnerships in this book illustrate a delightful range of experiences and share with the reader insights gained by the reserarchers and students involved in these projects. It gives teachers real examples of what happens when an exciting learning environment is created.

Gerry Wheeler
*Executive Director, National
Science Teachers Association*

Preface to the Series

The mandate to expand and improve science education for the 21st century is an educational imperative. Implementing change, however, is very complicated given that science education is dynamic, continually incorporating new ideas, practices, and procedures. Lacking clear paths for improvement, we can and should learn from the results of all types of science education, traditional as well as experimental. Thus, successful reform of science education requires careful orchestration of a number of factors. Technological developments, organizational issues, and teacher preparation and enhancement, as well as advances in the scientific disciplines themselves, must all be taken into account. The current prospects look bright given national reform movements such as the National Academy of Science's "Standards for Science Education" and the American Association for the Advancement of Science's "Benchmarks"; the backing of science education leadership; and recent developments, including the Internet and new educational software. Further, we have a world-wide citizenry more alert to the need for quality science education for all students, not just those who will become scientists. If we can isolate and combine such factors appropriately, we will have levers for science education reform. The books in this series deal in depth with these factors, these potential levers for science education reform.

In 1992, a multidisciplinary forum was launched for sharing the perspectives and research findings of the widest possible community of people involved in addressing the challenge. All who had something to share regarding impacts on science education were invited to contribute. This forum was the *Journal of Science Education and Technology*. Since the inception of the journal, many articles have highlighted relevant themes and topics: the role and importance of technology, organizational structure, human factors, legislation, philosophical and pedagogical movements, and advances in the scientific disciplines themselves. In addition, approaches to helping teachers learn about and use multimedia materials and the Internet have been reported. This series of volumes will treat in depth consistently recurring topics that can support and sustain the scientific education enterprise and be used to raise levels of scientific knowledge and involvement for all.

The first four volumes illustrate the variety and potential of these factors. *The Hidden Curriculum—Faculty-Made Tests in Science: Part 1, Lower Division Courses* and *The Hidden Curriculum—Faculty-Made Tests in Science: Part 2, Upper Division Courses* are premised on the belief that testing practices influence educational procedures and learning outcomes. Innovations in exam practices that assess scientific understanding in new and more appropriate ways should be shared with the widest possible audience. The research described and the resulting compendium of hundreds of contributed, annotated best exam practices in all science courses at the college level is a resource for every science educator and administrator.

Web-Teaching: A Guide to Designing Interactive Teaching for the World Wide Web aids instructors in developing and using interactive, multimedia educational materials on the World Wide Web. It also helps instructors organize and control these resources for their students' use. Not only do instructors learn how to improve their own materials and delivery, but they can access and make available Web-based information in a way their students can comprehend and master. Using the lever of instructional technology is an increasingly important part of science teaching; this book guides that process.

Finally, *Internet Links for Science Education: Student–Scientist Partnerships* illustrates the workings and effectiveness of this new paradigm and growing force in science education. In these partnerships (SSPs), students help scientists answer questions that could never before be fully addressed due to the lack of a large number of strategically positioned observers. Students gather and analyze data in projects involving authentic and important scientific questions, and science teachers actively explain science to students and help scientists implement their research. Data gathering and sharing, the heart of effective SSPs, is possible and rapid with the help of the Internet and a variety of technologies—groupware, visualization, imaging, and others. Several representative SSPs are described in depth. Chapters on student data and the human and technological infrastructures required to support SSPs help readers understand the interplay of the several factors in this approach to improving science education K–12. The Appendix contains a useful annotated list of current projects with complete contact information. Readers of this book will come away with an understanding of these programs from multiple perspectives and will be encouraged to become involved in similar efforts.

It is our hope that each book in the series will be a resource for those who are part of the science reform effort.

Karen C. Cohen
Cambridge, Massachusetts

Preface

This book explores the role of the Internet and technology in the rapid and successful development of a new form of science education and involvement, authentic student–scientist partnerships (SSPs). Its chapters illustrate the relatively recent development of this new field, present examples of varied projects and programs enabled or enhanced by the Internet and technological advances, and include information about implementation. Special chapters on student data acquisition, accuracy and verification, related technologies, and teacher enhancement efforts help to round out our understanding of the nature and scope of SSPs. The Appendix lists over 40 SSPs with annotations concerning student involvement in each and full contact information. This work is a resource for scientists, science educators, and science students, as well as educational administrators, parents, and community leaders as they become involved with this emerging, effective educational and science strategy. It is my hope that readers of this book will better understand various aspects of the SSP paradigm and that some will become actively involved with one or more programs.

The major considerations that helped me to structure this book were:

- The quality and improvement of science education both nationally and globally
- The appropriate use of technology in enhancing traditional and innovative aspects of science education
- The process of implementing change and improvement
- The development of human and technological communication channels required to support beneficial change

A stellar roster of authors produced its chapters in an extremely rapid time frame. The enthusiasm and help of these leaders in the SSP field and in its technological and human infrastructure are testimony to the timeliness of this work.

Numerous people must be recognized and thanked for their efforts on behalf of this book. First and foremost is Robert F. Tinker of the Concord Consortium. He involved me in the SSP field by asking me to coordinate a conference and edit its proceedings.[1] In these roles, I was uniquely positioned to learn about all aspects of this new force in science. Moreover, as an evaluator, involved with assessing the impact(s) of technology on educational programs, I naturally looked at this development in science education in terms of the varied and multiple factors creating a climate of support and facilitating its growth. This book is the result.

Bob Tinker wrote the first chapter, an overview of the SSP field. Dan Barstow and Sue Doubler of TERC similarly must be thanked; their cosponsorship with Bob Tinker of the Student & Scientist Partnership Conference reflects their deep involvement and prominence in the SSP field and Dan coauthored the chapter on Mars Exploration. Another key person is M. Patricia Morse of the National Science Foundation, the primary source of funding for the conference. Her boundless knowledge and enthusiasm supported much of the conceptual and project work reported here. She also coauthored the chapter on the NSF perspective.

Barrett N. Rock, first Chief Scientist of the GLOBE Program, was enormously helpful in developing this book. Not only did he appear to intuit needs before they were expressed, but he also created the chapter on GLOBE with colleagues, recruited additional authors, and found resources to produce a fine work. I am deeply appreciative of his efforts. All of the authors must be recognized for their professional knowledge and enthusiastic contributions. Their dedication produced wonderful writing, encompassing an enormous range of thinking and activity. Their help cannot be appreciated enough. In addition, everone who was asked wrote a chapter and did so on time! Oviously we share their enthusiasm for this book on such a timely topic.

Rachel A. Nixon wrote the chapter on EARTHWATCH, the first student–scientist collaborative project presented. In EARTHWATCH, the "students" are science teachers, working closely and intensively with scientists in research expeditions. These teachers, in turn, bring their new knowledge and creative excitement back to their own students in their home schools.

Rick Bonney and Andre Dhondt wrote the chapter on Project FeederWatch, for which they are mainstays. Their dual scientific and educational collaboration is so committed and involved they are an integral team in implementing their effective SSP program.

Forrest M. Mims III illustrates that amateur scientists doing relatively simple observations with networks of students can provide valuable information

---

[1] The Student & Scientist Partnership Conference was held in Washington, D.C., October 23–25, 1996. It was sponsored by TERC and the Concord Consortium and funded by the National Science Foundation. This conference brought together an invited group of 60 scientists, science educators, and supporters, formally to launch a new paradigm in science education, namely SSPs. The conference report details the rapid advance of SSPs along with issues and challenges in such partnerships for all parties and is available from TERC.

addressing complex questions. His chapter on the Sun Photometer Atmospheric Network (SPAN) and a variety of other scientist–student efforts well illustrates the ultimate efforts consistently implemented.

Ivar Babb, Peter Scheifele, and Domenic Tedeschi present an interesting and smaller project, the Aquanaut Program, which portrays both mentorship and partnership in science education. Heavily dependent on technology, this program covers a domain most projects cannot enter.

Dan Barstow and Cheick Diarra contribute a chapter on a new SSP, Mars Exploration. Still in its infancy, its focus and design show promise as a future SSP.

Russell G. Congalton and Mimi L. Becker explore the pivotal topic of student data quality in the GLOBE project and provide an exercise illustrating this concern and activity.

Alice B. Krueger, Patrick D. French, and Thomas G. Carter look at data acquisition, verification, and accuracy from the perspective of and provide examples of various simple and complex technologies in current use. Alice also developed the very current database of SSP resources and contact information in the Appendix. Her work should be valuable to all readers.

Three chapters focus on developing and implementing the technical and human infrastructure so necessary to implementing SSPs. M. Patricia Morse and Nora H. Sabelli of the National Science Foundation present the role and goals of NSF in supporting technology development, project development, and teacher enhancement programs.

Roy D. Pea, Louis M. Gomez, Daniel C. Edelson, Barry J. Fishman, Douglas N. Gordin, and D. Kevin O'Neill stimulate our thinking both about impacts and about the challenges of using a variety of technologies in collaborative efforts for all aspects of SSPs and other educational uses.

Richard Greenberg presents work on the development of one technology, image processing for teachers (IPT). The educational outreach efforts are explained in detail.

I also want to thank Betty B. Hoskins, Amelia McNamara, and Ken Howell for keeping me in-line and on time. Betty, both a professor of biology and skilled editor, made valuable contributions by clarifying and streamlining several of the chapters. Amelia McNamara and Ken Howell of Plenum Press encouraged both this book and my editing the entire series entitled Innovations in Science Education and Technology. Having conceptualized and edited this book myself, I now understand better what I have put the authors of other books in the series through. It's not so bad.

Happy reading!

Karen C. Cohen

Contents

New Pathways in Science Education—The Internet, Technology, and SSPs

Karen C. Cohen

A profound revolution is quietly and rapidly expanding the nature of science education throughout the world. Student–scientist collaborations have always been the hallmark of science education and inquiry at the forefront of scientific research. The traditional mentor and research assistant models, especially appropriate in laboratory and field settings, are expanding and extending through student–scientist partnerships (SSPs). Both technological advances and facilitation by science teachers are furthering this development worldwide. These collaborations, SSPs, have certain primary characteristics:

1. *Scientists* ask and use students to help answer questions that could never before be fully addressed because large numbers of strategically positioned observers were lacking.
2. *Students* gather and analyze data, so are involved in large-scale projects never before operable—projects that involve authentic and important scientific questions.
3. *Science teachers* are active intermediaries not only for explaining science, but also for helping scientists and students implement their research.

This book is particularly satisfying, because it combines a close look at SSPs with my experience in assessing the development and impact of a variety

Karen C. Cohen, Educational Technology Center, Harvard University, Graduate School of Education, Cambridge, MA 02138; cohenka@hugse1.harvard.edu or kccohen@mit.edu.

of educational reforms and my knowledge of the related technological and human infrastructures required to implement, support, sustain, and diffuse such innovations. These related factors include the Internet and a variety of technologies, the data acquisition and verification process, and teacher training or enhancement efforts.

Since the 1970s there have been global movements to involve scientists, science teachers, and their students in addressing major issues that cannot be studied with only a few scientists and small numbers of students. Chapter 1 presents an overview of the SSP movement. After defining and elaborating on the general SSP, it speculates on a variety of areas SSPs can address, and discusses required education and technological supports these efforts require and have. Chapters that fall within three major categories follow the overview:

- Selected Projects
- Student Data Acquisition, Accuracy, and Verification
- Enabling Technologies and Infrastructure Development

The Appendix contains an annotated directory of SSP partnerships with identifications and contact information as complete as we could find at the time this book went to press. It provides a perspective for the reader on current implementations and who to contact to become involved with these efforts.

The Selected Projects section covers a variety of student–scientist collaborations; each of which is vividly described. The act of selection, however, is idiosyncratic; the range of current SSPs and the disciplines in which they operate (and could operate) is by no means fully presented. Nonetheless, the selected projects cover the gamut of student–scientist collaborations and partnerships, and focus on the learning activities and the technologies involved

EARTHWATCH has been a major scientist–science teacher collaboration since the 1970s, impacting further on those teachers students. The GLOBE program represents most dramatically this type of very-large-scale SSP project, initiated in 1974 and currently involving scientists, teachers, and students in 43 countries Project FeederWatch has also been in operation since the 1970s, but the nature and type of scientific activities it can address has been dramatically altered in recent years by the Internet and communications technologies. There is also a long history of amateur contributions to furthering science and motivating students to become involved in science; thus SPAN (Sun Photometer Atmospheric Network) is presented in this book. The Aquanaut program is a small project in which scientists and science teachers work closely together and the science teachers, in turn, involve their students in authentic science research projects; it is heavily dependent on technology. Mars Exploration is similarly dependent on advanced technology. Although just beginning, it provides a framework for long-term science involvement, advancement, inquiry, and discovery, designed to include the best features of an SSP.

Large-scale SSP efforts involve or envision involving numerous scientists, hundreds of teachers, and thousands of students. The majority of such efforts have been advanced dramatically in the last few years primarily because modern technology—specifically the Internet infrastructure and developments in groupware, conferencing, visualization and imaging, data gathering and analysis, to mention but a few—has become a reality for many educational settings. We present rich descriptions of these selected projects to help the reader (and possible participant in these and future efforts) learn more about the exciting, rewarding, and varied opportunities in all forms of SSPs.

After these project pioneers have explained their programs and their uses of technology, we next present important chapters on Student Data Acquisition, Accuracy, and Verification with both complex and simple instruments.

> [A] significant challenge with student/scientist partnerships is data quality....The data are at the heart of a partnership, and the level of interest in access to student data by the scientific community will be determined in large part by the quality of the data. (James G. Lawless and Barrett N. Rock, in press)

The final section, Enabling Technologies and Infrastructure Development, has three chapters. An overview looks at the goals and role of the National Science Foundation in fostering the multitude of developments that further the new SSP paradigm. The next chapter covers a large number of networking technologies and requirements for putting them into educational use. Included are videoconferencing, structured collaborative groupware, and scientific visualization tools. The final chapter presents the development of a specific technology—image processing—and the support and teacher training necessary to implement the NSF and SSP goals.

Time and space have established the boundaries of the book. We hope there are no such limits on the future of these SSPs.

CHAPTER 1

Student–Scientist Partnerships
Shrewd Maneuvers

Robert F. Tinker

It is the very strangeness of nature that makes science engrossing, that keeps bright people at it, and that ought to be at the center of science teaching. I believe that the worst thing that has happened to science education is that the great fun has gone out of it....Very few see science as the high adventure it really is, the wildest of all explorations ever taken by human beings, the chance to catch close views of things never seen before, the shrewdest maneuver for discovering how the world works. (Thomas, 1981)

A NEW PARTNERSHIP

A new partnership is emerging between science and education based on the capacity of young learners to undertake studies, observations, and analyses of real value to science. Through participation in these student–scientist partnerships (SSPs), students gain a unique understanding of both the content and the process of science. The partnerships can enhance learning by supporting the most difficult aspects of the new science standards: providing increased opportunities for students to be engaged in extended inquiry. Conversely, SSPs can support a wide range of scientific studies by providing the capacity for worldwide observations, monitoring, and analysis by enthusiastic amateur student–scientists and their teachers. The rapid emergence of microcomputers and networks that are increasingly available in schools makes these partnerships feasible where they would have been unimaginable a few years ago.

Robert F. Tinker, The Concord Consortium, 37 Thoreau Street, Concord, MA 01742; bob@concord.org; http://www.concord.org.

The Potential of SSPs

The traditional way scientists support precollege education is often dependent on scientists donating time for an occasional lecture at local school or supervising a single student's advanced research project. This type of relationship rarely has a major impact on either students or scientists. SSPs, on the other hand, represent two-way streets of mutual interdependence and self-interest that have the capacity to grow and last. For some scientists, school-based research might be the best, most cost-effective way to gather data or undertake certain studies. Authentic science activities may bring to schools instrumentation, expertise, a sense of importance, and scientific "high adventure" that is otherwise inaccessible.

SSPs could grow to have the following impacts:

- Every student could have an authentic science experience at least once while in school.
- Every school could have at least one teacher-scientist involved in original studies, changing the school culture and infusing authentic science throughout a school.
- Scientists planning research projects could routinely use schools for large-scale data collection.

This chapter explores how SSPs could be expanded to reach levels of use that would significantly impact education and the conduct of science.

The Range of SSP Activities. Our attention is focused on a particular kind of partnership between students and scientists in which there is a *mutual dependency of scientists and students both engaged in the same studies*. Defined this way, SSPs can be a major force in the reform of education and have the ability for the relatively small number of available scientists to engage large numbers of students and significantly impact education.

It is important to note that SSPs can take place both inside and outside formal educational institutions. The Internet can engage individual learners anywhere and anytime and informal institutions such as museums can provide exactly the same kinds of SSP-based learning opportunities as schools. In fact, the nonschool application of SSPs could well be the only way to reach many learners who have either dropped out or are in schools without SSP programs. Thus, let it be understood that the term *student* as used in this essay, can apply to learners in both formal and informal learning settings.

For SSPs to reach all students, it is necessary to expand our definition of *scientist*. The term applies, of course, to the university researcher professionally engaged in funded research. But it can apply to anyone who needs student help in investigations. This could be a local conservation commission, a graduate student, an engineer, a bridge inspector, an amateur scientist, or even a precollege student; credentials are far less important than having an inquiring mind, knowledge of a topic, and an ongoing research program.

Teachers who are also researchers could become the most important means by which SSPs impact education. There are a few such teacher-researchers who pursue research programs and regularly engage their students in their own studies. For example, David Tucker offers physics and biochemistry research projects at Mt. Baker High School in Deming, Washington. Students are engaged in environmental research projects focusing on lead, zinc, phosphates, and nitrates in the soils, lakes, streams, and ground water that determine the health of salmon habitats. Students design valid investigations much like practicing scientists do and process their results similarly. The students also design protocols that other teachers can use to assess the health of their salmon habitats. This project, which is very technology intensive, has been funded by the State of Washington with technical support from the Concord Consortium. Each team of students works with university and agency scientists and collaborates with other students throughout the state doing similar projects. The partnership with scientists is active and two-way; both scientists and students benefit.

There seem to be four distinct models for SSPs:

1. *Scientist led.* Here the scientist defines the research program and protocols and students gather and analyze data. For example, the GLOBE project described later in this book involves a team of scientists each with NSF funding for a particular peer-reviewed study that involves scientist-generated protocols that students follow.

2. *Scientist guided.* In this case, the scientist defines and supports a topic of research, but students are involved in the design and conduct of the study. For instance, William Stapp is interested in river quality monitoring and has established the Global River Environmental Education Network (GREEN) to foster student water quality studies. GREEN publishes protocols, establishes how a single water quality figure of merit can be calculated, but leaves it to students to decide on sites, specific tests, and other features of their local study.

3. *Instrument based.* Here instrumentation needed by scientists is placed in schools but student and scientist uses of the instrumentation can be different. For instance, the Princeton Earth Physics Project (PEPP) at Princeton University places networked research seismographs in schools. The scientists are interested in the small earthquakes these instruments are uniquely able to observe while the students are directed to the more interesting and newsworthy major events. In this case, little of the student work is of scientific interest, but the mutual dependency is around the instrumentation and a common topic.

4. *Student originated.* Here the design of the experiment and the protocol used is largely determined by students while scientists are

available as consultants, advisors, or reviewers. For instance, the TERC Global Lab project (Berenfeld, 1994) has created a curriculum and activities to provide the knowledge, skills, and instrumentation students need to undertake their own investigations in certain areas. Students can find colleagues on the Net to undertake collaborative studies they define.

Of these four models, the scientist-directed is the most valuable to professional scientists whereas the student-originated studies are the most ambitious educationally; the other two models stake out intermediate positions between these two extremes. In all four models, the educational value depends critically on the skill of the teacher, the availability of related curriculum materials that address appropriate student learning goals, and the opportunities provided for active student participation in research.

There are many other kinds of activities scientists and engineers have created to bring research to precollege education. Some scientists invite students to observe the process of research with very limited student involvement. Some projects provide student internships in research settings while others have asked students to undertake studies where the result is known and, therefore, not of interest to the scientists. Other projects invite students to participate in a design challenge like that faced by scientists and engineers, but provide no substantial way for students to influence the actual design. Although these are generally useful projects, it is helpful to exclude them from the category of SSPs as they have little potential for large impact on education because they either cannot scale up or do not use effective educational strategies.

There is also a long, poorly documented tradition of high-caliber individual student research found in the best private schools, public science magnet schools, and schools of math and science, as exemplified by the Westinghouse awards. This tradition is important because it elevates research as the ultimate goal of precollege science achievement. It does not, however, meet our requirement for SSPs because it cannot reach all students and rarely is a true SSP.

A Strategy for National Impact

The mutual dependence of SSPs is the key to their growth. The excitement around the SSP concept is its potential to grow to a size that can transform education and have a major impact on science. Thus, it is important to see the relatively small projects that are appearing as only the beginning of a much larger effort. These and new projects need to be examined for their potential to expand and reach large size.

If SSPs were to reach every U.S. student once, they would have to engage about 5 million students annually. There are not enough scientists to get anywhere near this number if we were to rely on more traditional one-on-one scientist involvement. Even if there were 5 million scientists volunteering a bit of time to education, it would not have the impact on education that SSPs can. This is why it is important to be fairly strict about the definition of SSPs.

It is possible for SSPs to reach 5 million students annually. A project can easily involve 200 classrooms, or 5000 students. One thousand such projects would be able to reach all U.S. students, a large but feasible number of projects. It is more likely that a national strategy would involve a dozen huge projects like GLOBE and GREEN and hundreds more that would serve 50 or more classrooms each. Calculations such as these show that it is feasible to move the isolated, marginal SSP programs we currently have into the educational and scientific mainstream. A key to this is that self-interested collaboration is stable and scalable. Before looking at how this could be done, let us first understand why this would be so important to education.

THE EDUCATIONAL IMPORTANCE OF SSPS

The Importance of Student Inquiry

Genuine student inquiry and investigations are a foundation educators need on which to build student understanding of mathematics and science. It is only through their own inquiry that students learn the content and process of science. The most important challenge facing science education is to increase the utilization of student investigations. This requires that teachers understand that student investigation is essential and can only be guided by teachers who have had that experience themselves.

Unfortunately, this wildest of explorations has been squeezed out of most science education. In the rush to put more science content into science education, to prepare students for the next exam, the high adventure of science has been lost. Science education has developed into a separate entity divorced from science and research. From kindergarten through college, students rarely *do* math and science, they seldom see these topics as creative, and they do not view them as open-ended.

By ignoring real investigations, we not only fail to convey an accurate impression of what mathematics and science are, but we miss out on a teaching strategy that is fun, motivating, inspiring, and educationally sound. Students say this in their own words:

> Yeah, building solar houses, it was fun building them, cause we got to design our own and we got to record all like our materials and stuff. *We got to use our own imagination...to create.*

> They [the activities] are different, yeah, a lot different, 'cause I mean you have more hands-on and *there's no real boundary to what we can do.* (Wier, et al, 1992)

Supporting the New Standards

At this time, there is a major effort underway to strengthen U.S. science achievement by creating rigorous national standards that define what students are expected to know at different grade levels. Two similar national standards for science education have been promulgated, the Science Education Standards created by the National Research Council (NRC, 1996) and the Benchmarks for Science Literacy by the American Association for the Advancement of Science (AAAS, 1993). These national standards are expected to be interpreted by the states which will develop more detailed frameworks based on the national standards. In keeping with the decentralized nature of U.S. education, individual school districts will bear the responsibility for creating instructional strategies that adhere to the state frameworks and result in students meeting the national standards. This process sounds very top down, but it has been developed in a very inclusive manner that had generated widespread involvement and acceptance of these national standards. As this is being written, many U.S. states have created their own frameworks based on the national standards and schools are facing the challenge of changing their instruction accordingly.

The new science standards place "extended inquiry" at the very center of their recommendations. This is very difficult for schools, because it requires reconceptualization by teachers who are often unprepared in science and seldom experienced in research. This is even more daunting for most informal institutions because they are typically staffed by volunteers, parents, and staff that rotate quickly and seldom have research experience.

By place of honor and sheer bulk, the Benchmarks for Science Literacy places student inquiry front and center. In each section of the Benchmarks, inquiry is addressed first. The Benchmarks asks that students be able to "frame the question, design the approach, estimate the time and costs involved, calibrate the instruments, conduct trial runs, write a report, and finally, respond to criticism" (p. 9). A majority of the benchmarks are related to the skills students need to undertake and understand their own investigations and design projects; fewer than half of the benchmarks address traditional science and math content such as energy conservation, plate tectonics, and evolution.

Similarly, the National Science Education Standards (NRC, 1996) emphasizes inquiry: "For students to develop the abilities that characterize science as inquiry, they must actively participate in scientific investigation" (p. 173). The NRC definition of inquiry sounds a lot like research: "asking questions, planning and conducting investigations, using appropriate tools and techniques to gather data, thinking critically and logically about relationships between evidence and explanations, constructing and analyzing alternative explanations, and communicating scientific arguments" (p. 105).

Both sets of standards emphasize that the tools of inquiry are important to master because flexible tools are needed for student-initiated inquiry that might lead in directions and require measurements and analysis that are unanticipated. Information technologies are seen as particularly valuable tools; the standards mention the use of real-time data acquisition (MBL[1]), spreadsheets, data analysis and graphing tools, modeling software, electronics, and instrumentation.

Meeting the inquiry requirements is the most difficult challenge educators face in implementing the new science standards and causing the intended reform. In most schools, emphasizing inquiry requires changing the entire *tone* of science education. Science education must be changed from an emphasis on a huge array of facts and definitions supported by texts, lectures, and standardized labs. Because time must be found for inquiry, fewer facts and definitions will be "covered." Tests must be changed accordingly and parents and colleges educated to the importance of these changes. Because inquiry can lead in unexpected directions, more flexible resources and laboratory equipment are needed and teachers must be willing to follow student interests into unfamiliar topics. Teachers must metamorphose from experts with static knowledge in narrow areas into guides willing to learn about a much broader range of topics. In short, science education needs to become more like science.

SSPs can play a critically supporting role in fostering this change of tone by injecting real science into schools. We do not envision SSPs being the only source of inquiry-based activities, but by bringing authentic science into schools, they can become a driving force for needed change. If only one or two teachers in a school were actively involved in SSP research, they could have a major impact. The school would have to ask how best to prepare students for SSP activities and this would lead inevitably to increased use of inquiry. If the SSP teachers were teacher-researchers like David Tucker, then the attitudes, needs, and approaches of scientists would be more likely to color all of instruction and make inquiry part of the environment.

THE RESEARCH IMPORTANCE OF SSPs

There are many fields such as particle physics, bridge building, and theoretical chemistry that are better left to the professionals than to SSPs. This still leaves many research topics that could be part of SSPs.

A good SSP topic has a number of characteristics:

- Students and teachers can understand the science. There is little point in engaging students in incomprehensible research.
- The cost of suitable instrumentation is low. Assuming we want to reach large numbers of students, the equipment must not strain either the school's or the scientists' budgets.

---

[1] This acronym was coined by the author and stands for Microcomputer Based Labs.

- The research needs manpower or geographically distributed observers. There should not be other ways of doing the research more efficiently.

There are many kinds of investigations, projects, and activities that fit these characteristics and cannot be undertaken in other ways for a range of practical and cost issues.

One important class of potential SSP topics involves long-term studies. It is very difficult to recruit professional scientists into research that requires years of data taking; it sounds like a dead end career with limited opportunities for growth. This kind of research could be done at schools that represent stable sites with constant turnover of students who might enjoy making a solid contribution to long-term studies.

The following is a highly speculative list of possibilities for SSP research offered simply to illustrate the possible range topics that fit the required characteristics:

- *Environmental studies.* Studying the environment has been quite popular because the science is accessible and there are unlimited opportunities for important local studies. Although many water, soil, air, and acid rain studies have been developed and intense, long-term studies of a particular site have been undertaken, there is the need to disseminate what has been done and to develop a wider range of environmental SSP projects.
- *Large-scale biodiversity studies.* How are we going to find out what the current state of the environment is in terms of the diversity of species? Answering that question requires a huge amount of observation and investigation. The costs of undertaking a national biodiversity inventory are prohibitive unless SSPs are used.
- *Epidemiology.* There are many epidemiological studies of low level diseases that could be undertaken, e.g., the common cold. These are not reportable illnesses, but better information about propagation would be valuable.
- *Meteorology.* At first blush, meteorology seems ideal for SSPs until you realize that existing meteorological monitoring is so extensive that students can add little, except where their observations are unique. Examples of the latter are: quantifying atmospheric haze, doing micro-climate studies, and making observations, such as in the West Pacific, where other reports are not available.
- *Ozone and UV.* There are surprisingly few long-term records of stratospheric ozone or ground-level ultraviolet levels. Inexpensive instruments for measuring these have been developed that are suitable for student use, but need to be employed in scientist-led research.
- *Image analysis.* There are huge banks of images that students could analyze. There are, for instance, pictures of Mars that no one has ever analyzed. Astronomy could use more eyes to spot supernovas and other

transitory events. As we shall see later in the book, sophisticated software for image analysis is beginning to be used in schools.

- *Sociocultural studies.* Involving students in data gathering and analysis of social and cultural issues is a relatively unexplored area. Students could tell us volumes about living conditions in various subcultures, about drugs, and about peer pressures to which they are subjected. Being part of the studies would give them access to information and techniques to resist antisocial behavior.
- *Educational research.* There is a need for more information about educational practice and impacts. We do not know, for instance, the magnitude of the suspected drop in laboratory experimentation in schools. Students could be invaluable in tracking the large-scale impact of technology and school reform while they acquire important information about cognition and educational strategies.
- *Instruments.* The PEPP seismographs might be one representative of a class of instruments that could be built and maintained by schools and used in different ways by students and researchers. Automated weather stations, monitoring systems for nuclear plants, and water quality monitoring equipment could all contribute to SSP projects.

TECHNOLOGY AS FACILITATOR

As noted earlier, there is a history of bringing research into schools in various ways. Beginning at the turn of the century, extended student investigations were an integral part of the progressive education movement that were largely buried by the urgent demands of World War II but persist to this day. Although these earlier efforts have been important, they represent a marginal part of education.

Our optimism for massive application of SSPs in education at this time is based on the current revolution in information technologies, specifically electronic instrumentation, microcomputers, and telecommunications. These technologies are beginning to create options for SSPs that would not have been possible a decade ago. These technologies make it feasible to support investigations involving large numbers of students all around the world in collaboration with scientists who do not have to travel to all of those places in order to make a significant contribution. Let us examine the most important of these technologies and the contribution each can make to SSPs.

Collaboration

The Internet and networking provides critical support for partnerships. Of particular interest are asynchronous forms of communication that remove the scheduling problems of telephones and electronic conferencing. The conflicting schedules of schools and scientists mean that asynchronous communication is essential for SSPs.

The richness of asynchronous communication is just now increasing. Where we have been previously limited to text-based electronic bulletin boards or conferences, now business needs for collaborative technologies are making asynchronous collaboration much more flexible. The documents that are exchanged now can contain graphs, pictures, line drawings, and voice as well as text. In addition, the documents that a collaborating group creates can be viewed in different ways and worked on by the group to create a body of knowledge that complements individual understandings.

Data Acquisition

We can now attach sensors to inexpensive computers that allow all kinds of measurements and investigations based on flexible instrumentation that only a few years ago graduate students would have yearned for. Embedded controllers today use RISC processors that can contain a complete, powerful computer in an eight-pin integrated circuit costing a dollar. Soon, these controllers will execute Java at the silicon level. At the same time, sensors are being continually improved and integrated with electronics. For instance, modern inexpensive light and temperature sensors contain their own amplifiers and calibrations so that they generate linear outputs that are easy to incorporate in other circuits. These components are not only driving down the costs of circuits, they also open the possibility, because few components are necessary, that students could assemble sophisticated instrumentation from kits.

These developments will create the possibility of a new generation of low-cost smart sensors and interfaces that could serve data anywhere on the Internet. Attached to wireless hand-held computers students could take these sensors anywhere to measure data accurately and effortlessly. This can reduce the drudgery for students and help reassure scientists of the accuracy of data.

Data Sharing

An important part of a functioning SSP is the ability to quickly and accurately share data among large numbers of students and researchers. Not only should the data flow from students to scientists, but students should also be able to see the combined data as soon as possible. Over a decade ago, I developed the first network-based data sharing software for SSPs as part of the National Geographic Kids Network. Now, because databases are the key to many business network applications, there is a range of Internet-based technologies that routinely support data sharing. Two attractive approaches are to use Tango to serve Filemaker Pro files and to use Lotus Development's Notes which supports distributed databases.

Data Analysis

It is an important part of the SSP learning experience to engage students in analyzing the data they collect. This is why we added data mapping and graphing functions to the earliest NGS Kids Network software and its next generation called Alice. The range of data analysis tools applications that could be used in schools is now huge. As Java and other modular code becomes more widely supported, basic data analysis functions will become more widely available and less expensive.

The four enabling technologies described above are all just taking off, but they are sufficiently mature now to support far wider use of SSPs; in the next few years they will only get better, and both schools and scientists will become more comfortable using them to support SSPs.

NEXT STEPS

Partnerships between educators and scientists around school-based research represent a new phenomena, often enabled by technology. These partnerships are poised for large-scale expansion that could greatly improve education and give science valuable new resources.

There is a growing group of experienced and innovative educators and scientists who have experience in various aspects of SSPs. However, the field has not yet developed a literature because so many of the projects are action-oriented and have little time or professional incentives for publication. As a result, the knowledge of project structure, supporting technology, school implementations, and student impact is not being recorded or shared. Further, new projects typically make predictable mistakes that could be avoided. Before the field can expand, it is important to codify what has been learned and to use this experience to develop plans that could expand the reach of SSPs.

We need to begin a decade-long effort to consolidate our knowledge, address the barriers, and scale this effort up to reach all students. This could well result in projects to create an infrastructure of curriculum, technology, and expert groups that will support large scale implementation of SSPs that could reach all students. The goal of this book is to begin this process.

REFERENCES

American Association for the Advancement of Science, *Benchmarks for Science Literacy*. (New York: Oxford University Press, 1993).
B. Berenfeld, "Technology and the new model for science education: The Global Lab experience," *Machine-Mediated Learning*. 4(2/3) (1994): 203–227.

National Research Council, *National Science Education Standards* (Washington, DC: National Academy Press, 1996). See also <http://www.nap.edu/readingroom/books/nses/>.

L. Thomas, *Humanities and science,* Presented at the Sloan Foundation's "Conference on new dimensions of liberal education," Key Biscayne, Florida (New York: Alfred P. Sloan Foundation, 1981).

S. Weir, "Electronic communities of learners: fact or fiction?" in R. Tinker and P. Kapisovsky, eds., *Prospects for Educational Telecomputing: Selected Readings* (Cambridge, MA: TERC, 1992).

CHAPTER 2

The GLOBE Program
A Model for International
Environmental Education

Barrett N. Rock, Tommie R. Blackwell,
Danielle Miller, and Anne Hardison

INTRODUCTION

The GLOBE (**G**lobal **L**earning and **O**bservations to **B**enefit the **E**nvironment) Program is a hands-on, international environmental science and education program that creates a partnership between students, their teachers, and the scientific research community. The students participating in GLOBE make measurements of selected atmospheric, hydrologic, and biologic parameters, following protocols developed by the research community, which, in turn, uses these student-collected data sets in their own research.

The GLOBE Program was formally proposed on Earth Day, April 22, 1994, by Vice President Al Gore. At present, teachers from over 3800 schools in the United States and 50 countries have been trained to present GLOBE materials to their students. Over 100 nations have expressed interest in participating in this exciting and innovative environmental education program.

The goals of the GLOBE Program are to:

- Enhance environmental awareness of individuals worldwide
- Increase scientific understanding of the Earth
- Support higher levels of achievement in science and math

Barrett N. Rock, Complex Systems Research Center, University of New Hampshire, Durham, NH 03824; barry.rock@unh.edu.   Tommie R. Blackwell, Director of Education, The Space Camp Program, U.S. Space and Rocket Center, Huntsville, AL 35807; tommieb@spacecamp.com. Danielle Miller and Anne Hardison, The GLOBE Program, 744 Jackson Place, NW, Washington, DC 20503; dmiller@globe.gov and hardison@globe.gov.

GLOBE brings children, teachers, and scientists from around the world together in an integrated science/education effort focused on studying Earth. The science component of GLOBE is designed to meet two objectives: first, to introduce students to hands-on scientific measurement and analysis, and second, to provide the research community with access to sources of accurate and meaningful data.

The educational component of GLOBE is designed to introduce students to the process of doing "real science." GLOBE allows students to "learn by doing" through a series of hands-on science and education activities designed to make them active participants rather than passive observers. Learning to do science is like learning to play basketball; doing it is better than reading about it. Being active participants allows students to "learn by doing" and to know they are involved in and contributing to important processes and products.

Through GLOBE, we all benefit: scientists, students, and teachers. Noel Brown, of the United Nations Environment Program (UNEP) in New York, states that "on Spaceship Earth there can be no passengers, only members of the crew." The GLOBE Program is designed to make participating students members of the crew.

A series of scientifically sound measurements are carefully designed to be appropriate to the student's age and grade level. Students learn not only how to carry out measurements but also to analyze their own data and data from other GLOBE schools. Student data sets are sent, via the Internet and World Wide Web (WWW), to GLOBE processing centers, where they are converted into visualization images and posted daily on the Web GLOBE home page (*www.globe.gov*).

Participating GLOBE scientists are actively involved in two ways: they analyze and validate the student data, and they actively inform the students how their data are being used and what interpretations are being made. All members of the scientific community will have access to GLOBE student-generated data via the Web. In many cases these student data sets will provide basic information unavailable in other ways.

Students from kindergarten through 12th grade are active participants in the GLOBE Program through their schools. Participation is designed to be age-appropriate for grades K–5, 6–8, and 9–12 (or equivalent worldwide). Younger students may make simple measurements which may be qualitative rather than quantitative. Older students will make additional, more sophisticated measurements as appropriate for their grade level.

THE GLOBE PROGRAM AND EARTH SYSTEMS SCIENCE

The study of environmental science has led to an understanding that everything is connected to everything else. Once humankind ventured into space, the truth of this statement was underlined. Studying the Earth from orbit reminds us that our planet consists of a network of interconnected systems, and that changes in one system will result in changes in other systems. From a student's standpoint, such a concept may be intimidating and processes may seem too complicated or too massive to be understood. So, GLOBE divides three Earth systems (Atmosphere, Hydrosphere, and Biosphere) into small, easily understood pieces. The various GLOBE measurements and learning activities are made at one of three study sites located near their school: the atmosphere study site, the hydrology study site, and the biology study site. These measurements, made over time, will allow the student to see on a local scale how the Earth's systems are interconnected in a dynamic and predictable way, producing an increased knowledge and understanding of how our planet works. (See Table 1 for a list of GLOBE Measurements.)

In addition, the view from orbit allows the students to see that their school and its surrounding areas are parts of a vastly larger region. It provides them with a "big picture," leading to a better appreciation for the global view they will see in the GLOBE visualization images. To help the students develop

Table 1. Scientific Measurements to Be Made by GLOBE Students and the Equipment to Be Used

Core measurements	Equipment used
Atmosphere	
Air temperature (maximum/minimum and current)	Maximum/minimum thermometer
Precipitation (liquid and solid)	Rain gauge
Cloud cover	Eyes
Cloud type	Cloud chart
Hydrosphere	
Water temperature	Thermometer
Dissolved oxygen	Dissolved oxygen kit
Water pH	Indicator paper/pen/meter (varies with student grade)
Alkalinity	Alkalinity kit
Electrical conductivity	Conductivity meter
Biosphere	
Land cover	Landsat data, *MultiSpec* software
Tree height/circumference	Clinometer/tape measure
Canopy/ground cover	Densimeter
Grass biomass	Scissors, tape measure
Species identification	Dichotomous keys
Phenology (seasonal patterns)	Camera
Soil moisture	Sample cans, auger, oven
Soil characterization	Color chart, graduated cylinder
Global Positioning System (GPS)	
Latitude/longitude of study sites	GPS receiver

a "view from orbit," each participating GLOBE school is provided with a Landsat image of their own region, covering an area approximately 15 × 15 km in size at 30 m pixel resolution. Weather satellite images produced from NOAA AVHRR (Advanced Very High Resolution Radiometer) data are also provided. The AVHRR images cover a much wider area, with 1-km pixels. Schools with the appropriate computer capabilities are provided with digital versions of the satellite data, along with freeware image processing software. Finally, daily temperature data collected by students will be returned the following day via the Web, in the form of a global-scale GLOBE Visualization Image. In this manner, students come to have an improved understanding of how the planet works at a range of scales, from local to global.

Common assumptions are made that the land surface of our home planet has been accurately mapped, and that the surface conditions are well known. However, according to Dr. Jack Estes, an earth scientist at the University of California at Santa Barbara, only 33% of the Earth's land surface has been accurately mapped for the distribution and location of features such as rivers, lakes, reservoirs, land cover types (e.g., forest types, wetlands, urban areas), and surface conditions, at a scale of 1:25,000 or greater (Estes and Mooneyhan, 1994). Many scientific research projects require an accurate understanding of surface conditions at such a scale (1:25,000), especially if such projects study issues on a global scale. Local studies rely on researchers who have access to their study areas and who can accurately measure surface conditions of interest; but for global-scale studies, such access is not always possible. For many areas of the world, accurate measurements do not exist of even such simple parameters as air temperature and precipitation. But climate models need accurate measurements on a global scale of simple parameters as well as complex surface features such as land cover type.

A group of international environmental scientists selected the initial GLOBE environmental parameters that—because they will be measured by GLOBE schools all over the Earth—will contribute significantly to the scientific understanding of the global environment. GLOBE schools worldwide make a core set of environmental measurements in the three Earth systems: Atmosphere (weather/climate), Hydrosphere (hydrology/water chemistry), and Biosphere (biology/soils). Each GLOBE school is encouraged to coordinate its activities with those of other neighboring GLOBE schools. As the GLOBE Program develops, some GLOBE schools will make elective measurements of local interest to selected schools.

Students can be *the* experts regarding the details of their local environment. Because students live in these areas, they can become an extremely valuable source of accurate data for use by the research community if they are properly trained as to what data are needed and how to objectively and accurately measure the environmental features of interest. In the process of understanding the issues being addressed by scientists and making the measurements suggested, the students are introduced to the process of doing real science. They experience a sense of ownership of the data produced and of contributing to improved scientific understanding of our home planet. Vice President Gore has suggested that the daily visualization images produced from GLOBE student data, combined with data from other sources, will represent the most accurate picture of Earth available (Gore, 1994).

Scientists use GLOBE student data in a variety of ways. Some will use GLOBE information about the actual conditions in their local area. Some will compare this ground validation information with predictions based on satellite data for these same areas. Satellite images of surface features must be calibrated against actual measurements of these features; if properly trained, students will be an excellent source of such calibration data. Other researchers will use the GLOBE data as input for their global models or as a means of testing the accuracy of their models' predicted results. Some will use the student data to more accurately map and monitor environmental conditions for large areas of the world.

Scientists will make use of the student-generated data in their ongoing research activities, only if they can be certain that the data sets are accurate. The scientific protocols provided by GLOBE are designed so that, if properly followed, data will be accurate. In addition, most of the equipment used in making the GLOBE measurements can be calibrated, another way to improve the accuracy of the measurements. Finally, GLOBE schools are likely to occur in clusters within an area (multiple schools in the same city) and will allow scientists to average student measurements, thus improving accuracy. Thus, the Vice President's hope can become a reality, and GLOBE data furnished by students will provide the scientific community with access to the most current and accurate picture available of Earth.

The scientific integrity of the GLOBE Program has been assured by scientists from each relevant discipline who develop the measurement procedures and the supporting educational materials, and who assess data accuracy. In this way, GLOBE scientific measurement procedures are accurately presented; broader educational materials are scientifically valid, and quality control is applied throughout the acquisition and processing of GLOBE data. Scientists join with educators in GLOBE Science/Education Teams for each scientific discipline area. Each team oversees all activities in its area and participates in the instruction of GLOBE students.

THE GLOBE PROGRAM AND EDUCATION

The GLOBE Program combines the use of environmental educational materials with the opportunity for students to conduct protocol-directed hands-on science. In the first year of GLOBE (1994–95), environmental educators and curriculum development specialists developed GLOBE educational materials, where possible adapting them from existing environmental education materials. In the second year, new educational materials and additional science protocols were developed by scientist/educator teams. In both cases, GLOBE materials are designed to be support materials used with existing curricula, rather than a new curriculum. Teachers participating in the GLOBE Program receive an intensive introduction to both the science protocols and associated educational materials, as well as implementation strategies. This requires attending multiple-day regional GLOBE workshops.

The GLOBE educational materials include a GLOBE teacher's guide with age-appropriate materials. It contains procedures for the GLOBE environmental measurements, broad information about environmental topics, and student work sheets and data entry pages which assist measurements and data reporting. Other environmental education materials that may be useful (including information accessible through the Internet) are identified for participating schools.

Country Coordinators and teachers from GLOBE partner countries (see Table 2) receive training at international regional workshops. Once these GLOBE-trained teachers and Country Coordinators have completed the workshop, they are able to both teach their own students, as well as train other GLOBE teachers in their respective Countries.

All of the major GLOBE student measurements involve learn-by-doing, hands-on activities designed to expose students to methods and equipment used by research scientists. In Fig. 1, GLOBE students from the Czech Republic use a densimeter to measure and record the percent canopy closure characterizing their 30 × 30-m pixel-sized biology study site. They will make multiple canopy measurements along the diagonals of their "pixel," following protocols developed by NASA research scientists involved in assessing ground conditions and relating the conditions to satellite image data.

Table 2. Countries that Have Joined the GLOBE Program as of April 1997

Argentina	Egypt	Kazakhstan	Romania
Australia	El Salvador	Korea, South	Russia
Austria	Estonia	Kyrgyzstan	Senegal
Belgium	Fiji	Luxembourg	South Africa
Benin	Finland	Marshall Islands	Sweden
Bolivia	The Gambia	Mexico	Tanzania
Canada	Germany	Moldova	Trinidad and Tobago
Chad	Greece	Morocco	Tunisia
China	Ireland	Netherlands	Turkey
Costa Rica	Israel	Norway	United Kingdom
Croatia	Italy	Palau	United States
Czech Republic	Japan	Poland	Uruguay
Ecuador	Jordan	Portugal	

Figure 1. GLOBE students from the Czech Republic use a homemade densimeter (modified toilet paper tube) to measure percent canopy closure for their pixel-sized biology study site. Canopy closure estimates provide one measure of biomass for a given land cover type.

Once trained by their teacher, GLOBE students will make daily temperature measurements, record them on GLOBE student work sheets, and enter them on the GLOBE WWW data entry page, to be sent to the NOAA Forecast Systems Laboratory in Boulder, Colorado. Using a National Weather Service Forecast Model, GLOBE student data contour maps of temperature (maximum and minimum) are produced and posted on the GLOBE home page the following day, along with maps of student-monitored precipitation and cloud cover.

GLOBE STUDENT VISUALIZATIONS

GLOBE visualization images are produced using GLOBE student data sets. Research scientists located at the NOAA Forecast Systems Laboratory in Boulder, Colorado, and the NASA Goddard Space Flight Center in Greenbelt, Maryland, within 1 day incorporate the daily student-gathered data into global visualization images of maximum and minimum temperature, precipitation, cloud cover, and soil moisture. These images let GLOBE students see how their data sets have contributed to the "big picture." They are able to compare their measurements, made on the previous day, with the same daily measurements made by other GLOBE students in their own area, in their respective countries, and in all of the participating GLOBE countries.

Figure 2a presents a student data contour map, a GLOBE visualization image of maximum temperatures reported on December 20, 1995, by approximately 75 participating GLOBE schools in the United States. The "+" marks on the map indicate the locations of the reporting GLOBE schools. This image was available on the WWW GLOBE home page on December 21, 1995. The "Nor'easter of 1995" was in full force during the hours of GLOBE data collection (± 1 hour of solar noon) on December 20, and the coldest color patterns (maximum temperatures below 0°C) coincide with areas of maximum snowfall.

Students also are able to compare their ground measurements with "reference data sets," similar measurements derived from satellite data or model predictions. Such comparisons allow students to see that satellite-derived data and/or model predictions may not coincide with their measurements (their "ground truth"). This is because the satellite may be measuring the actual temperature of the ground, rather than the air temperature 1.5 m above the ground which is the location of the GLOBE weather station. Temperature measurements made by several GLOBE schools in a relatively small area, may reflect local variations resulting from differences in local land cover type (urban versus rural, open versus forested) as well as topography and exposure. A weather satellite averages temperature for areas greater than 1 km². In addition, the thermal signal reaching the satellite has to first pass through the atmosphere. What the satellite actually "sees" of the ground varies as the different air masses move through an area, and atmospheric conditions change.

GLOBE students also can compare predicted values presented in their daily newspaper with their data. Figure 2b presents the National Weather Service's predicted (modeled) temperatures for the United States, for 1 P.M. EST, for the same day as Fig. 2a (*Boston Globe*, December 20, 1995). A comparison of the GLOBE student data contour map (Fig. 2a) and the modeled predictions (Fig. 2b) reveals some very strong correlations. However, the GLOBE data are in degrees Celsius, and the National Weather Service predictions are in degrees Fahrenheit, so the exercise requires a mathematical conversion process. A math lesson is incorporated into the activity. Such comparisons of student-derived

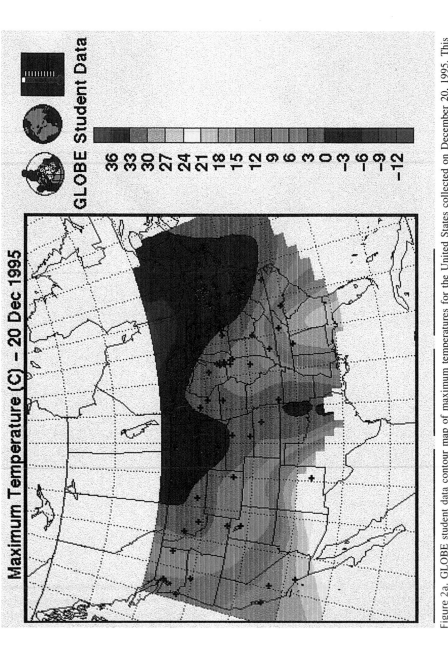

Figure 2a. GLOBE student data contour map of maximum temperatures for the United States collected on December 20, 1995. This example of a GLOBE visualization image appeared as part of the GLOBE home page on the World Wide Web (www.globe.gov) on December

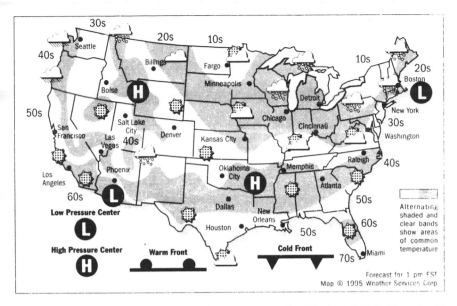

Figure 2b. National Weather Service map of temperatures predicted for 1 P.M. EST for December 20, 1995, as they appeared in the local newspaper. Note that the GLOBE student data are real data (maximum temperatures for the previous 24 hours) whereas National Weather Service data represent a modeled product.

(real) data with modeled (forecast) data allow students to better understand that models may or may not represent reality. GLOBE student snowfall data (not shown in Fig. 2a) for the week of December 18 allowed students to track the actual path of the storm.

Some measurements (water temperature, pH, and related chemistry) are made weekly, and the visualization images also will be produced weekly. The biometry measurements (vegetation height, stem diameter, canopy closure, ground cover, and species identification) and land cover type measurements will be made yearly, and the visualization images of these environmental parameters will be generated once a year. All of these visualizations will be provided to students and interested research scientists via the Web.

Some measurements may have no appropriate reference data set. In other cases, the reference data set may be a modeled product that has no direct equivalent to a student measurement product. One of the student-generated products will be a land cover map, for the 15 × 15-km Landsat image of the area near each GLOBE school. Land cover maps are designed to provide researchers with spatial patterns, which represent the kinds of vegetated and nonvegetated ground-cover types characterizing each area. The dominant land cover type for the students' 30 × 30-m biology site will be used to characterize or describe most of the GLOBE biometry measurements, or measures of biomass (vegetation height, stem diameter, percent canopy closure/ground cover), along with species identification. This biology site is within a larger 90 × 90-m land cover site, and near other cover types typical of the students' local area. Such land cover maps may be produced by visual interpretation of a hard copy Landsat satellite image or by computer, using a freeware image processing software called *MultiSpec* . The hard-copy image, digital Landsat data, and *MultiSpec* are provided to domestic schools by GLOBE. Such student-generated land cover maps will assist scientists in characterizing and validating larger, regional cover types. Both the extent and type of land cover for a given area have a direct impact on the amount of solar radiation reflected back to the atmosphere, but some regions of the planet are data-sparse. Such validation data will be useful in improving climate models currently being developed.

Other scientists can make use of the visualization images in much the same way. Comparing the student's data sets with the reference data sets, they can gain a better understanding of exact relationships between "ground truth" data and the products of satellite assessments or model predictions. In many cases, the GLOBE data represent the first time the scientific community will have access to such "ground truth" for many areas of the globe.

GLOBE COMPUTER AND COMMUNICATIONS SYSTEMS

The GLOBE Program builds on the rapidly expanding Internet infrastructure throughout the world. The easy-to-use WWW multi-media information-access capability is the basis for GLOBE communications, including data entry and school-to-school communications. Domestically, each GLOBE school has an IBM-compatible or Apple Macintosh computer, suitably configured and connected to the Internet with WWW access. Students use readily available WWW Internet software to enter their data and transmit it to a GLOBE data processing center.

International GLOBE schools are also encouraged to use the same approach. Initially, some partner countries' GLOBE schools may not have computers or access to the Internet. Then the nation's GLOBE Country Coordinator is responsible for reporting measurement data to the GLOBE data processing site via the Internet. A variety of media, including photographs and broadcast television, may be used to distribute GLOBE environmental visualization images to such schools. GLOBE will accommodate international partners with whatever level of computer and communications technology they bring to participation in the program.

The ability to communicate among GLOBE schools has inspired exciting exchanges of ideas and information. For example, a Spokane, Washington, school initiated an Internet-based project with GLOBE schools in Wisconsin, Korea, and Sweden to share information on atmospheric trace gases. A GLOBE school in Australia initiated a GLOBEMail relationship with a school in Kentucky; the Australian school principal and GLOBE teacher visiting their colleagues in the United States. A GLOBE school in New Hampshire used the Internet to arrange a trip for seven of its students to visit several GLOBE schools in the Czech Republic. While in the Czech Republic, the students described to the Czech students how they were using GIS (Geographic Information Systems) and *MultiSpec* to manage and map land cover types on their school's campus.

IMPLEMENTING THE GLOBE STRATEGY

GLOBE received the first student data on the 25th Earth Day, April 22, 1995, with over 1500 schools in the United States and overseas participating. Thousands of additional schools throughout the world have joined GLOBE, with the number of schools growing as rapidly as resources are available domestically and internationally. As of April 1997, a total of 51 countries have joined the GLOBE Program (Table 2).

Teachers introduce students to the GLOBE Program. In an intensive, three-to four-day workshop, each GLOBE teacher receives training in conducting the measurement protocols, utilizing the GLOBE learning activities, and navigating the GLOBE Web site on the Internet. The workshops introduce teachers to the GLOBE teacher's guide, the hands-on measurement and learning activities, and the details of how and when to make the measurements. A portion of each workshop invites the teachers to discuss, in small, grade-appropriate groups, how best to introduce GLOBE into their classrooms.

The overall approach is to provide teachers with materials that supplement their existing curricula. Materials are presented, not as yet one more thing to be added to an already full plate, but rather as a hands-on activity to be used in place of current laboratory or lecture projects. GLOBE teachers are given the flexibility to integrate the GLOBE activities into their curricula as they find most appropriate. Some GLOBE schools have developed a coordinated program in which different classes make different sets of measurements. Some schools have the younger students make simple measurements (e.g., pH of water using litmus paper) while the older students in those schools make more sophisticated measurements of the same parameter (e.g., pH of water using a pH pen or meter).

GLOBE is designed to instill responsibility for environmental stewardship among students and, ultimately, all peoples of the Earth. Domestically, the program will be inclusive. Internationally, the broadest possible participation from geographically diverse regions of the world is sought. This makes both educational and scientific good sense, as many data-sparse regions of the world are the locations of participating GLOBE schools.

Abraham Funchess, a GLOBE teacher in Matthews, South Carolina, reported, "This is cross-curricula teaching at its best. It gets the whole brain involved in the thinking process. We are having so much fun with our GLOBE Program and environmental studies that school simply does not end for us at 3:00 P.M. The GLOBE Program is serving as the magnet that attracts boys and girls to being champions of the environment."

Ms. Dana Votapkova, the GLOBE Coordinator in the Czech Republic, states, "The measurements made by the children in GLOBE are not only for themselves, but also will be used, once they are evaluated, by the scientists of the world. They will also tell their parents and their community about what they learn."

"In my 23 years of teaching, I have never seen such positive student reaction to a new science program," says GLOBE teacher Bob Jost of Fresno, California.

GLOBE IN ACTION

Since the GLOBE Program received the first student data in April 1995, extensive student and teacher participation has been limited by a lack of on-line capabilities in many of the schools in both the United States and abroad. As more schools come on line, and more countries join, the Program will provide an ever-expanding network of students with the opportunity to increase their understanding of our planet, and will supply access to meaningful data for a growing number of members of the research community. Two examples of recent activities will be cited that highlight the effectiveness of the GLOBE Program for both students and research scientists.

The first example is an e-mail sent by GLOBE students at Broken Hill High School (BHHS) in Australia to the over 30 other GLOBE schools in Australia.

>Return-Path: <bhhs@opennet.net.au>

>X-Sender: bhhs@opennet.net.au (Unverified)

>Date: Thu, 7 Dec 1995 13:58:02 +0930

>To: ausglobe@erin.gov.au

>From: bhhs@opennet.net.au (BROKEN HILL HIGH

SCHOOL)

>Subject: Hello from the GLOBE students at BHHS

>G'Day everyone!!

>

>This is the first time that we the students at BHHS have used

the INTERNET

>& the mail service. We think that GLOBE is really exciting and

>interesting. We have learnt so much already - we think our brains will

>explode (some people don't seem to have them).

>

>Hope things are going good at other schools, and you are hav-ing as much fun

>as us! Please email us soon to tell us how you are going.

>

>All the best,

>

>Sarah, Megan, Ed, Nick, Kate, Brylie, Chantelle, Nichole, Louise, Susan,

>Roslyn, Amy, Rebecca, Erin and Mr. Langfield.

The second example serves to illustrate how eager researchers at the state level are to have access to student-generated data. Ms. Judy Silverberg, of the New Hampshire Fish and Game Division, used the GLOBE home page to contact all 29 GLOBE schools in her state. She requested access to their maximum/mini-mum temperature data and their snow depth data for the winter of 1995–96. New Hampshire was developing a Winter Severity Index, assessing the impact of low temperatures and heavy snowpack on the survival rate in the state's white-tailed deer population.

White-tailed deer may be killed by low temperatures when their access to normal winter browse (typically twigs and bark of maple, hobblebush, eastern hemlock, and balsam fir trees) is restricted by deep snow. Once snow depth reaches a critical level it is difficult to move out of their deeryards. Limited mobility leads to limited access to good browse, which in turn makes the deer more susceptible to death at low temperatures. Young deer (fawns) are the most susceptible because of their high energy demands, and their lower capacity for heat retention as measured in their higher ratio of body surface to volume.

The temperature and snow depth data gathered by GLOBE students will be used to predict survival rate and determine the critical snow depths for specific regions of New Hampshire. In addition, these data will be fed to a Fish and Game computer system that will monitor critical levels and trigger a state-level response, providing supplemental natural browse to endangered areas. Researchers in the Wildlife Management Program at the University of New Hampshire are identifying the most important types of natural browse for use in deployments in response to local conditions. According to the New Hampshire State Fish and Game Department, the addition of the GLOBE school data to the existing statewide database is a significant contribution to the development of the Winter Severity Index.

GLOBE MANAGEMENT

In the United States, the GLOBE Program is managed as an interagency effort led by the National Oceanic and Atmospheric Administration (NOAA). The other federal agencies involved in the program are the National Aeronautics and Space Administration (NASA), the National Science Foundation (NSF), the Environmental Protection Agency (EPA), the Departments of Education and State, and the Council on Environmental Quality (CEQ) as well as the Office of Science and Technology Policy (OSTP) in the Executive Office of the President. The nonprofit Foundation for Global Environmental Education raises private sector support to complement the efforts of the federal GLOBE Program. GLOBE international partners are responsible for implementing their own GLOBE activities.

GLOBE schools worldwide receive functional and performance specifications for the required instruments and lists of possible vendors and instrument models. Before their measurements are regarded as accurate and reliable, GLOBE instruments must meet these specifications; they need not, however, be purchased from identified vendors.

JOINING GLOBE

All K–12 schools in the United States are invited to join GLOBE. Non-U.S. schools should contact their GLOBE Country Coordinator. For more information on the GLOBE Program or GLOBE Teacher Training Workshops, please contact the GLOBE Help Desk at 1–800–858–9947 or send email to info@globe.gov. The mailing address is The GLOBE Program, 744 Jackson Place, NW, Washington, DC 20503.

REFERENCES

J.E. Estes and D.W. Mooneyhan, "Of maps and myths," *Photogrammetric Engineering and Remote Sensing* 60(1994):517–524.
A. Gore, Personal Communication. Meeting with GLOBE Staff, November 2, 1994, Vice President's Ceremonial Office, Old Executive Office Bldg., Washington, DC.

CHAPTER 3

FeederWatch
An Example of a Student–Scientist Partnership

Rick Bonney and André A. Dhondt

INTRODUCTION

As the 20th century draws to a close, educational research is painting a new vision of science education. In this vision, articulated by the National Science Education Standards (National Academy of Sciences 1996), science is best taught and learned when students engage in extended inquiry and research. "Inquiry into authentic questions, generated from student experiences, is the central strategy for teaching science," the Standards asserts. "Students should experience science that engages them in actively constructing ideas and explanations and that enhances their abilities to do science."

One excellent way for students to experience real science in the classroom is to conduct research in ornithology, a field to which nonprofessionals have been making contributions for decades. Indeed, from the early 1800s, when a Scottish poet, Alexander Wilson, published the first comprehensive account of North American birdlife, to the turn of the century, when the Audubon Christmas Bird Counts were born, to the present time, when thousands of bird watchers participate in bird censuses, surveys, and research projects across the continent, amateurs have gathered a treasure house of information on bird numbers, bird movements, and even basic bird biology (Bonney, 1991).

These amateur-assisted projects offer many opportunities for science learning. Consider Project FeederWatch, a survey of winter birds conducted by

Rick Bonney, and André A. Dhondt, Cornell Laboratory of Ornithology, 159 Sapsucker Woods Road, Ithaca, NY 14850; reb5@cornell.edu and aad4@cornell.edu.

volunteers across North America, which began in 1987 as a joint project of the Cornell Laboratory of Ornithology (CLO) in the United States and the Long Point Bird Observatory in Canada. FeederWatch was started with two goals: (1) to collect data about continental bird populations which can be obtained only by large numbers of volunteers spread across a large geographic area and (2) to teach participants about birds and ornithological study by involving them in the scientific process (Bonney, 1988, 1994).

Nine years later, the project has shown demonstrable success. During the winter of 1995–96 more than 11,000 people were enrolled, representing every U.S. state and every Canadian province. More than 500,000 data forms have been submitted, and trends in winter bird populations are beginning to emerge. FeederWatchers also have conducted experiments to determine what kinds of seeds birds like best, and have helped to track the spread of an eye disease (conjunctivitis) that is infecting house finch populations in the eastern United States. Results from Project FeederWatch have appeared in many newspapers and magazines (e.g., *Birder's World*, 1995), and are starting to appear in the scientific literature (e.g., Wells *et al.*, 1996).

Project FeederWatch also has been successful from an educational perspective. Evaluations of the project have been conducted through focus groups, telephone interviews, and mail surveys. All have shown that participants are improving their knowledge about birds and the process of science (Hezel Associates, 1992; Cornell Office of Communication Strategies, 1994, 1995).

To reach its full potential as a tool for science education, however, FeederWatch must address some important issues. We will elaborate on these later, but to summarize: First, the audience of current participants—older and highly educated—is probably not the audience most in need of science education. Second, present communication between participants and professional scientists is mostly one-way, so a true "partnership" between amateurs and professionals has not yet been established. Third, feedback to participants concerning findings from their data presently takes several months.

FeederWatch also faces an important problem as a tool for research: At present, CLO biologists must discard from 5 to 10% of the data submitted each year because they are incorrectly submitted or the information seems implausible.

CLO, in collaboration with TERC and the Cornell Theory Center, is addressing these issues through a two-step process. First, to reach a younger, more formative audience and to establish true student–scientist partnerships, we are developing a version of FeederWatch called Classroom FeederWatch, designed for upper elementary and middle school students, and operated over the Internet.

Second, to improve feedback to participants and the quality of data submission in the original Project FeederWatch, we are developing an Internet version that can be used by anyone connected to the World Wide Web.

THE FEEDERWATCH PROCESS

The design of Project FeederWatch has remained simple and essentially unchanged over the years. Participants are solicited from the CLO membership and the general public through direct mail, advertisements, news releases, and targeted mailings to bird clubs and chapters of the National Audubon Society. They receive a "research kit" including an instruction booklet, tally sheet, computer-scannable data forms, and a data form return envelope. During winter they follow a standardized protocol, carefully explained in the instruction booklet, to record the highest number of individuals of each bird species that visits their feeder on one or two, self-selected, "count days" every 2 weeks. In spring they send their data forms back to the CLO, where biologists analyze them and write a report on the season's feeder bird numbers, distributions, and movements for the CLO newsletter, *Birdscope*. Later, the data are combined with information from previous years to determine population trends.

SCIENTIFIC RESULTS

The simple count data collected by FeederWatch participants can be used in many ways. Much can be learned in the course of a single season. For instance, changes in the population size of a species can be tracked throughout the winter; or populations of a species can be compared among different geographic regions. Comparing data from many winters, long-term population trends can be discerned in both bird numbers and distributions. And, FeederWatchers can engage in studies or experiments of feeder bird biology. Some specific examples follow.

Within-Season Population Change

An example of population change within one winter is illustrated by the graph in Fig. 1, which shows the percentage of feeders visited by Carolina wrens in the Southeast and Northeast regions of North America, throughout the winter of 1993–94. In the Southeast region the number of wrens remained fairly constant throughout the winter, but in the Northeast region, Carolina wrens visited fewer and fewer feeders as the season progressed. The decline began after a severe storm that dropped at least 3 feet of snow over most of the region (Tessaglia and Rosenberg, 1994).

Within-Season Geographic Variation

An example of geographic variation in bird social organization within one winter is illustrated in Fig. 2, an index of black-capped chickadee group sizes at feeders in five states and one province, also throughout the winter of 1993–94.

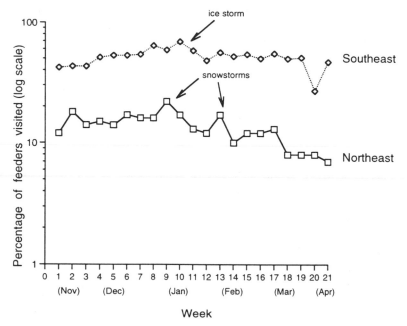

Figure 1. The percentage of feeders visited by Carolina wrens in the Southeast and Northeast regions of North America, throughout the winter of 1993–94, shows how Project FeederWatch can document bird population changes during the course of just one winter.

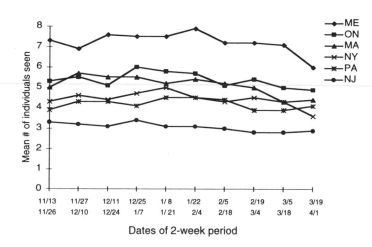

Figure 2. An index of black-capped chickadee group size, measured in five states and one province throughout the winter of 1993–94, shows how Project FeederWatch can document geographic variation in bird social organization throughout a single season.

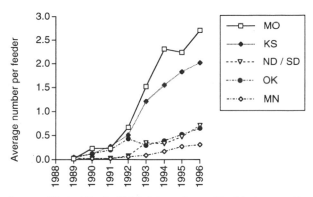

Figure 3. Several years of Project FeederWatch data can be combined to document long-term trends in populations of birds such as the house finch. This graph shows how the species is increasing rapidly in the Great Plains, where it was virtually absent less than one decade ago.

Two findings are apparent from this graph: First, group size remained stable in each region over the entire FeederWatch season. Second, group size varied among different regions, perhaps reflecting variable foraging strategies in different climates (Dhondt and Lowe 1995).

Long-Term Trends

Combining data from several years, an example of a long-term population trend is illustrated in Fig. 3, which shows the average number of house finches seen per feeder in the Great Plains from 1989 through 1996. This species, originally restricted to the western United States, was introduced in New York in 1942. From there it has been spreading back toward the Rocky Mountains, where it may eventually meet up with the original western population (Tessaglia and Rosenberg, 1996).

Additional Feeder Studies

In addition to counting birds, some FeederWatchers have engaged in experiments such as the national Seed Preference Test conducted during the winter of 1993–94, and in special studies such as the House Finch Disease Survey, begun in 1994 and still ongoing.

The Seed Preference Test (SPT) was one of three "National Science Experiments" funded by the Informal Science Education program of the National Science Foundation. The experiments were intended to involve the public in a series of guided activities that would answer specific ornithological questions while teaching participants about birds and the process of science. The SPT was

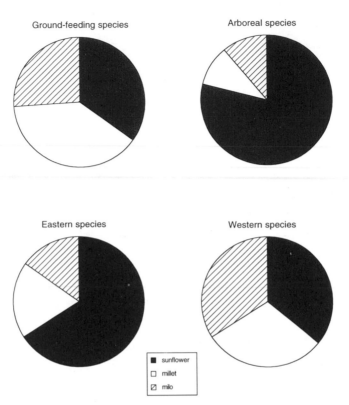

Figure 4. Data from the Seed Preference Test clearly showed birds' preferences for three types of seeds: black-oil sunflower, white proso millet, and red milo. These pie charts show average, national seed preferences of four contrasting groups of birds: 18 species of ground feeders including sparrows, towhees, doves, and cardinals; 15 species of arboreal feeders including chickadees, nuthatches, jays, and finches; 8 species restricted to the eastern United States and Canada; and 8 species restricted to the Southwest and the West Coast.

designed to answer the question, "What types of seeds do birds like best?" The 7000 FeederWatchers who elected to participate received a supplemental research kit that included an SPT instruction booklet, computer-scannable data forms, and a color poster of common feeder birds. (About 10,000 non-Feeder-Watchers solicited from the general public also participated in this project.)

In the spring of 1994 the CLO received data from about half of the participants, who had carefully followed an experimental procedure to test birds' preferences for three types of seeds: black-oil sunflower, white proso millet, and red milo, also known as sorghum. These data clearly showed seed preferences for many species of North American birds (Fig. 4). The number of birds that eat milo—nearly one-third of the western species—was particularly interesting,

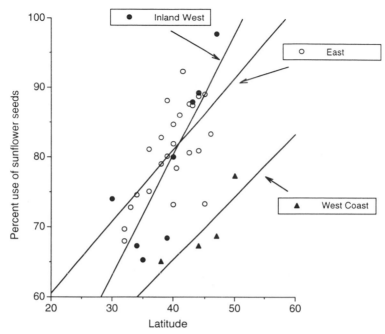

Figure 5. The Seed Preference Test also showed that the amount of black-oil sunflower seeds consumed by several bird species, including the house finch, varied with latitude. House finches living in the North ate more sunflower seeds than those living in the South. Each point on the graph represents data from a single state or province.

because conventional wisdom had stated that birds don't eat these seeds (Rosenberg and Bonney, 1994). Even more interesting, however, are regional seed preferences. For instance, the amount of sunflower seeds eaten by house finches at different latitudes is shown in Fig. 5. Not only do house finches at northern latitudes consume more sunflower seeds—a high energy food source—than those in the South, but finches on the West Coast ate fewer sunflower seeds than would be predicted by their latitude, perhaps because the climate is milder on the coast than in the mountains (Rosenberg 1996).

The House Finch Disease Survey was begun in 1994 to track the spread of an eye disease caused by the bacterium *Mycoplasma gallisepticum*. FeederWatchers who elected to participate received yet another supplemental research kit, again consisting of simple instructions and data forms. These folks have been observing whether house finches visiting their feeders are healthy or sick. Participants have sent in more than 22,000 monthly data forms, which have clearly shown the spread of the disease (Fig. 6). When the first data arrived in November 1994, they showed that the disease was already well established in

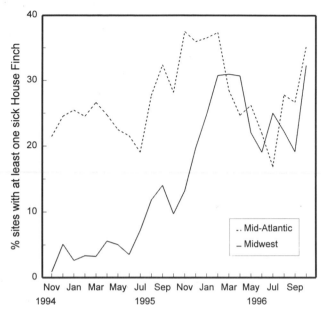

Figure 6. The House Finch Disease Survey has documented changes in the incidence of eye infection within the eastern U.S. house finch population. When the survey began in November 1994, it showed that the disease was already well established in the Mid-Atlantic, from Vermont and New Hampshire south to Maryland. During the winter of 1995–96, the disease spread into the Midwest, and is now found in all states east of the Mississippi River.

the Mid-Atlantic from Vermont and New Hampshire south to Maryland. During the winter of 1995–96, the disease spread into the Midwest, and is now found in all states east of the Mississippi River (Dhondt, 1997).

ACCURACY OF DATA

Accuracy of data is an issue for this project, indeed, for any citizen-science project. Each year project staff must reject from 5 to 10% of the data sent in, either because the data forms are unreadable (smudges or missing information), or because the data seem implausible (i.e., many more individuals of a species reported at a given location than experience has shown to be likely).

EDUCATIONAL RESULTS

FeederWatchers' knowledge of birds and understanding of the process of science has been documented in several ways. First, the FeederWatch Data Form Booklet includes a comment form, and each season, thousands of these are returned to the CLO where they are read, categorized, discussed, and if necessary, answered. Overwhelmingly, participants mention the fun and enjoyment they receive from the project; frequently they mention what they are learning. For instance, on the 1996 comment forms, a participant from Westford, Massachusetts, wrote, "Wonderful project—I've learned a lot about the birds I thought I knew." A participant from Naples, California, wrote, "I love doing the count—it has made me much more aware of the birds' world." One from Fredonia, Arizona, said, "My observations have become fine tuned, and my enjoyment of bird watching has increased incredibly!" Many participants also note the enjoyment they get from contributing to science. A FeederWatcher from Augusta, Georgia, wrote, "I enjoyed watching my bird feeders having a special purpose. It was a good feeling knowing that I was contributing to a process that would help the birds." And one from Athens, Maine, stated "I feel like I'm actually doing something to benefit the natural world besides just giving money."

In addition to this anecdotal information, we have conducted several formal evaluations of FeederWatch during the last few years. These have included five focus groups with a total of 44 participants (Hezel Associates, 1992; Cornell Office of Communication Strategies, 1994); one survey mailed to 600 participants (Cornell Office of Communication Strategies, 1995); and two telephone surveys of 121 participants, including 101 FeederWatchers in 32 states who had not renewed their participation from the previous year (Hezel Associates, 1992; Cornell Office of Communication Strategies, 1995).

While FeederWatchers participating in these evaluations have been enthusiastic about the project, they also have offered many suggestions for improvement. Their ideas have led to continuous refinements in the project, including simplifying the data forms; expanding and clarifying the instructions, especially the protocol for counting birds; including more information in the instruction booklet about the value of the data to scientists; and increasing the number of project newsletters from two to four each year.

Our most recent survey was conducted in the fall of 1995, when our mailed questionnaires resulted in 317 written replies for a response rate of 52.8% (Cornell Office of Communication Strategies, 1995). Several findings are presented below.

Demographics

FeederWatchers are an older group, with 36% over the age of 65, 33% between 51 and 65, 27% between 36 and 50, 4% between 21 and 35, and only 1 respondent (~0%) under the age of 21 (Fig. 7).

FeederWatchers are also highly educated: 31% possess an advanced degree, 33% have a bachelor's degree, and 8% have an associate's degree (Fig. 8).

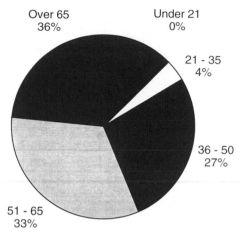

Figure 7. FeederWatch: Age of participants.

Project Materials

Respondents rated the Instruction Booklet very highly (4.7 on a scale of 1 to 5, with 5 being excellent) in terms of clarity of writing; appearance; usefulness and organization of content; and illustrations (Fig. 9). Their favorite sections were (1) the Instructions summary, (2) Instructions on filling out the data forms, (3) What's new for the upcoming season, and (4) How to make observations.

Respondents also found the Data Booklet easy to use, rating it an average of 4.6 on a scale of 1 to 5, with 5 being excellent (Fig. 10).

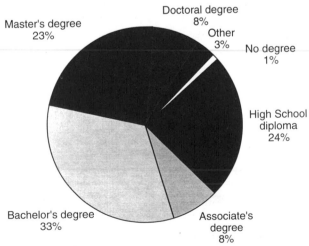

Figure 8. FeederWatch: Level of education.

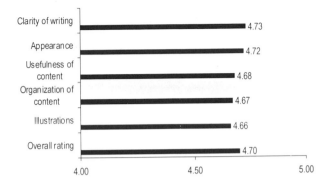

Figure 9. FeederWatch Instruction Booklet: Rating on a scale of 1–5, with 5 being excellent.

Learning about Science

Respondents reported that by participating in FeederWatch they learned a lot about scientific observation, data-gathering techniques, and bird behavior. Almost 60% of respondents stated that they learned a lot about bird behavior, with an additional 35% learning some (Fig. 11); 31% learned a lot about scientific observation, with an additional 56% learning some (Fig. 12); and 38% learned a lot about data-gathering techniques, with an additional 50% learning some (Fig. 13). Respondents with lower levels of education reported being more likely to feel that they learned a lot about scientific observation, data-gathering techniques, and bird behavior.

Feedback to Participants

Birdscope is read by 91% of respondents, and the "FeederWatch News" section was rated highly, with clarity of writing scoring 4.5, and usefulness of

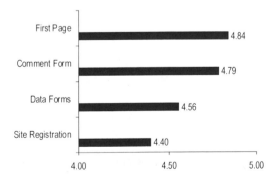

Figure 10. FeederWatch Data Booklet: Rating on a scale of 1–5, with 5 being excellent.

Internet Links for Science Education

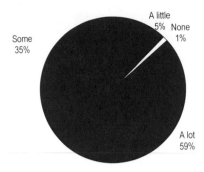

Figure 11. FeederWatch participant learning: Bird behavior.

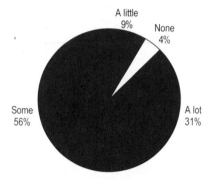

Figure 12. FeederWatch participant learning: Scientific observation.

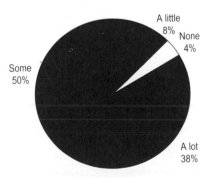

Figure 13. FeederWatch participant learning: Data-gathering techniques.

the information scoring 4.3 (both on scales of 1 to 5, with 5 being excellent). And, more than 96% said that a sufficient amount of information is provided on how their data are being used.

Concerns about feedback, however, are evident from some of the additional comments that respondents wrote on their surveys, such as: "More detailed relationships between FeederWatch data and professional research information might be useful," "Would like to see more information on species and localities, rare sightings," "Wish even more data on more species could be included," "Would like to see more data—we gather a lot more than seems to be used."

Concerns about feedback also emerged from our focus groups. Summarizing two sessions held in June, 1994, the evaluators wrote

> Participants overwhelmingly expressed the need to receive more feedback from the CLO regarding the project. Many said they felt they were becoming less interested in sending in data because they were uninformed about what the results are. Reporting in *Birdscope*, while interesting, seems inadequate. Many said they want to know more than just counts; instead they would like to learn about conclusions and implications of their findings. Many also said they would like to be able to access reports on their specific areas. And, several participants requested that the newsletter become more active in connecting FeederWatchers, for instance, by printing more letters and providing specific stories about watchers' experiences. (Cornell Office of Communication Strategies 1994)

Summarizing the effectiveness of FeederWatch as a tool for science education, we believe that it is on the right track, but that it faces four key problems.

First, FeederWatch as presently operated probably is not reaching the audience that stands to gain the most from participation, that is, younger and less educated people. We would like to be reaching a much larger audience, particularly of young people who are not already interested in birds.

Second, existing communication between participants and professional scientists is mostly one-way. That is, the professionals instruct the participants in data collection procedures, accept their data, analyze them, and publish results in the newsletter, but the participants have limited opportunity to work with and understand the data themselves, or to inform the investigations through their own thinking. Thus, a true "partnership" between amateurs and professionals has only partly been established. We would like FeederWatch to be truly a two-way communication, so that participants with questions, suggestions, or ideas for analyzing data or adding to our studies have easy access to the ears of the staff. Also, we want participants to be able to access the data for their own use and analyses, and to have access to scientific publications resulting from the project.

Third, a lag of several months exists between the time that participants send in their data (April) and the time they receive the season's results in the fall newsletter (September). In between their data are edited, scanned, and analyzed. The time lag is particularly troublesome for teachers trying to use Project

FeederWatch as an educational tool in their classrooms: By the time results using the students' winter data are published the following autumn, the children are already in a new class, most likely with a new teacher.

Fourth, we have to throw out some of the data because of smudging, partial entries, and implausible results. We want to improve the quality of data submission so that we can use and analyze all of the data we receive.

We are now addressing these issues through a two-step process. First, to expand our audience and to build a true student–scientist partnership, we are developing Classroom FeederWatch (CFW), a multidisciplinary, middle-school curriculum. It is keyed to the National Science Education Standards, and will be delivered over the Internet. This project will reach a larger and younger audience, will allow instantaneous and personalized feedback to participants, and will allow participants to access and analyze FeederWatch data themselves.

Second, to enhance the original project by improving the amateur–professional partnership, accuracy of data submitted, and feedback to participants, we are building an Internet version of Project FeederWatch, using the tools, techniques, and database management procedures being developed by CFW.

CFW OVERVIEW

CFW is being developed by the CLO and TERC with funding from the Instructional Materials Development program of the National Science Foundation. Like Project FeederWatch, CFW allows students to participate in the longstanding tradition of amateur-assisted bird study. But it goes much further. By connecting students and CLO scientists over the Internet, it allows for a true student–scientist partnership. Students receive continuous updates of data and findings from the project, and they ask questions and seek advice from the on-line scientists. The project also encourages students and their teachers to seek answers to their own questions, including some whose answers are presently unknown. Through CFW, students develop content knowledge about birds; learn about the iterative nature of scientific research; experience the importance of accurate communication of data and findings; and learn the basics of data analysis. Interdisciplinary connections include math, art, geography, language arts, social studies, and computer science.

CFW has four underlying principles:

1. Students are amateur ornithologists.
2. Students contribute data to a research database.
3. Professional ornithologists use student-generated data in their studies.
4. Students share their data over the Internet with other student ornithologists, analyze those data, and use their findings to describe how the natural world works.

The project consists of five or six class periods of Advance Preparation and 15 Explorations of several class periods each, conducted over the 9-month school year. Each Exploration includes an overview, interdisciplinary connections, list of purposes, main concepts to be covered, preplanning activities, an introduction, the activity, teaching tips, ideas for discussion, homework, ideas for assessment, and preparation needed for the next Exploration. The 15 Explorations are divided into four parts, each of which emphasizes a different aspect of bird studies.

Part I: Students conduct preliminary research. Students learn about the CLO's interest in receiving data from Classroom FeederWatchers, and they begin their study of birds by posing questions about them. Then they use a standard procedure to identify birds that visit a feeder outside their classroom window. They venture into the schoolyard to record field data about the birds they see and hear. They also select a research partner—a group of fellow Classroom Feeder-Watchers at another school—and write an introductory electronic letter in which they tell about themselves and their research. These initial Explorations result in some answers and in many questions to explore.

Part II: Students learn about birds and the environment. As students consider birds as an essential part of the natural world, they collect information about the characteristics of their FeederWatch site, and electronically send it to the CLO for later use. They design and build feeders to accommodate the needs of birds. And, they continue communication with their research partner. They also have the opportunity to set up experiments to probe their own questions.

Part III: Students learn about bird biology and behavior. Students explore beaks and feet, feathers, and bird digestion to discover how structure complements function. They consider the benefits of different feeding behaviors. They follow the FeederWatch protocol to count the kinds and numbers of birds that visit their feeders, and they transmit their data to the CLO over the Internet, using on-line data forms at the project Web site. They also have more opportunities to probe their own questions.

Part IV: Students use their project findings to learn more about birds. By downloading the data they have collected throughout the year, as well as data collected by other classes across the country, students analyze the accumulated CFW bird data. They make comparisons, search for patterns, make and evaluate predictions, discuss implications, and draw conclusions. They also design and conduct an in-depth investigation of their own, present their findings to an outside audience, and ask new questions. They create a class newsletter, *Classroom Birdscope,* which they share in both hard copy and over the Internet with members of their school, their research partners, and the CLO. They also select one feature of their newsletter to send to the CLO for incorporation into a national edition.

Throughout the project, students and teachers receive help from CLO staff in three ways. First, all classes are connected through a listserve on which participants share questions, findings, and tips for attracting birds and conducting research. CLO staff monitor this listserve and add to the conversation where appropriate. Participants can also ask questions by sending e-mail directly to the CFW electronic mailbox at the CLO. Finally, a huge amount of resource information is available, and is being continuously updated, at the CLO Web site (http://www.ornith.cornell.edu). This material is available to anyone with Web access, and includes information ranging from bird feeding to technical ornithology.

At the end of the school year, students discover what the CLO has learned from their collected data, as CLO biologists complete preliminary analyses and post them to the Web site.

This project would not be possible without the Internet, and a computer with Web access is essential for full participation. However, the computer is viewed as a tool for research, not an end in itself. Students spend the majority of their time conducting hands-on experiments and discussing their investigations, both in small groups and as a whole class. The computer is used to write letters, to send and retrieve data, to make tables of experimental results, and to create graphs to discover patterns and trends in the data. But the emphasis of CFW is to connect students with their environment—to get them looking past their computer screens to their schoolyard habitat and beyond.

The CFW curriculum includes several tools designed to help teachers assess what their students are learning as they work. These assessments also help teachers plan their projects, enhance their students' learning, and tailor their teaching to meet their students' needs. The project includes the following assessment tools:

- Pre-Project Questionnaire
- Post-Project Questionnaire
- Daily Assessments: Homework and Class Discussions
- Three Key Assignments
- An Embedded Assessment

These tools were designed by curriculum experts at TERC according to five central principles:

1. Assessment tools should help students consolidate, synthesize, and demonstrate their understanding of key learning goals. The CFW tools are scheduled to be used at critical stages in the project to help students pull their ideas together before they proceed to new material.
2. Assessment tools should motivate students to go beyond the information given to them. The CFW tools present a problem or question that requires students to apply key concepts and to use a range of talents to explain, design, or produce material on their own.

3. Assessment tools should require students to demonstrate their own knowledge to an audience. In CFW, students present their findings to a learning community including their own classmates, their research partner, and scientists at the CLO.
4. Assessment tools should be embedded in the curriculum. The CFW tools help students master key concepts as they move from Exploration to Exploration, and help them understand how each Exploration relates to the framework of the entire project.
5. Assessment tools should provide useful information. The CFW tools reveal what students understand and what they don't. Teachers can use this information to plan Explorations, to shape discussions, and to grade students on their progress. These tools also help students keep track of their own progress.

CFW also supports the National Science Education Standards. Exactly how the project activities line up with the Standards is shown in Fig. 14.

CFW PROGRESS TO DATE

The first draft of the CFW curriculum was written in the fall of 1995 and pilot tested with eight classrooms near Boston, Massachusetts, and Ithaca, New York, in the winter of 1995–96. It was revised during the spring and summer of 1996 and is being field tested in about 60 classrooms nationwide over the winter of 1996–97. At the time of writing, very few bird count data have been collected, so little can be said about the quantity or quality of the data that students will contribute. We do know, however, that birds are coming to classroom feeders, that classrooms are successfully connecting over the Internet, and that classrooms are successfully completing the introductory explorations.

Says one field test teacher: "This is a wonderful experience. I'm excited and feel very lucky to be part of it! This is the best use of telecommunications." Another teacher says: "I am really excited about this project. There are so many teachers looking for ways to make student learning more meaningful, and this is a wonderful project that fills that need." Still another states: "Thanks so much for reaching out to our kids. This program brings science alive and truly has the power to change lives."

We are also encouraged by the first batch of questions that the students have proposed for their own research, many of which are likely to result in interesting findings. "This is my favorite part of the project," says one field test teacher in Florida. "When we start to brainstorm ideas in class, my students get really excited about exploring their own questions."

Some of the questions we've seen include: Do birds take food back to their homes? Do birds like bright colors? How much do birds eat? Do different birds

Start-Up Activities

Start-Up Activity 1: Put a Feeder Outside a Window

Start-Up Activity 2: Register Your School

Start-Up Activity 3: Set Up Your Research Center

Start-Up Activity 4: Administer the Pre-Project Questionnaire

Explorations Part I: Preliminary Research

Exploration 1: What is *Classroom FeederWatch?*

Exploration 2: Which Birds Visit Our Feeder?

Exploration 3: What Can We Learn From Our Field Studies?

Exploration 4: What Will We Write to Our Research Partner?

Explorations Part II: Birds and the Environment

Exploration 5: What Is a Bird?

Exploration 6: What Data About Our Count Area Will We Send to the Lab?

Exploration 7: How Will Our Feeders Look?

Exploration 8: What Can We Learn From Our Research Partner?

Explorations Part III: Bird Biology and Behavior

Exploration 9: What Do We Count at Our Feeder?

Exploration 10: How Do Beaks and Feet Help Birds Eat?

Exploration 11: What Do Birds Do to Get Food?

Exploration 12: How Do Feathers Work?

Explorations Part IV: Continuing to Learn About Birds

Exploration 13: What Can We Learn From Our Bird Data?

Exploration 14: What Can We Learn From Our Questions?

Exploration 15: What Will We Share In Our Newsletter?

Year-End Activities

Year-End Activity 1: What Did the Lab Learn From Our Data?

Year-End Activity 2: What Did We Learn?

Figure 14. Classroom FeederWatch alignment with the National Science Standards.

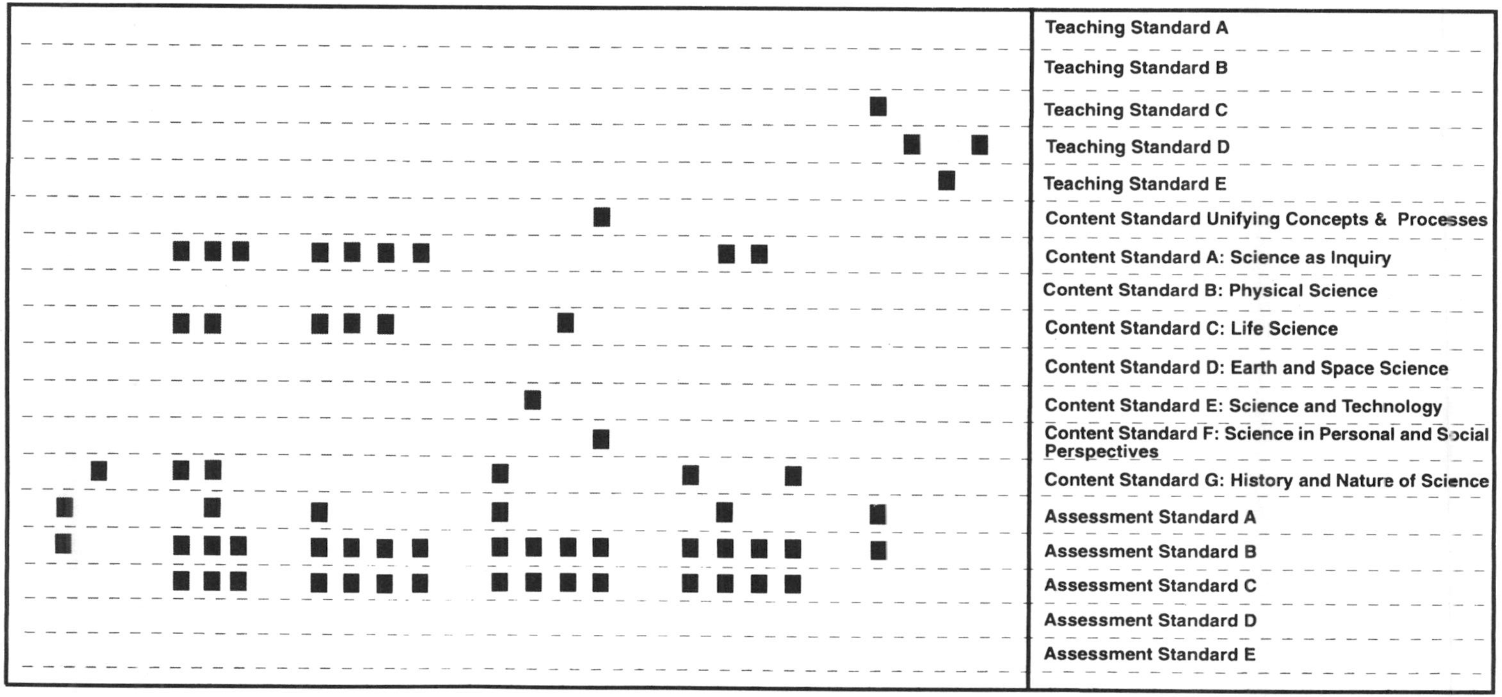

Teaching Standard A

Teaching Standard B

Teaching Standard C

Teaching Standard D

Teaching Standard E

Content Standard Unifying Concepts & Processes

Content Standard A: Science as Inquiry

Content Standard B: Physical Science

Content Standard C: Life Science

Content Standard D: Earth and Space Science

Content Standard E: Science and Technology

Content Standard F: Science in Personal and Social Perspectives

Content Standard G: History and Nature of Science

Assessment Standard A

Assessment Standard B

Assessment Standard C

Assessment Standard D

Assessment Standard E

Figure 14. (*Continued*)

eat different foods? All of these are questions that the students should be able to answer by making careful observations through their classroom windows.

Of course, we've heard constructive criticism as well, mostly about the time it takes to send and receive e-mail, about improper use of the listserve, and about difficulties in selecting research partners over the Internet. All of this feedback will be used to revise the curriculum as we proceed with development.

Formative evaluation of CFW is under the direction of Bruce Lewenstein, Associate Professor of Communication and of Science and Technology Studies at Cornell University. Lewenstein is working with a team of graduate and undergraduate students as well as the CFW staff to conduct classroom observations, teacher interviews, student pre- and posttests, and analyses of questionnaires that teachers complete on line at the end of each section of the project. A final, summative evaluation will be conducted by an outside contractor.

THE FUTURE OF CFW

We expect a final field test of invited classrooms to take place during the winter of 1997–98, and that the completed curriculum will be available to any classroom in the fall of 1998. CFW will be published and managed by the CLO, using Web software for data submission, retrieval, and analysis developed by the Cornell Theory Center.

One feature of the data submission process being developed by the Theory Center is automatic, real-time data editing. In the first field test, any data that students submit are accepted by the computer housing the database, but by 1997, all data will be screened as they come in. Participants will receive a display either showing the data they have submitted and thanking them for doing so, or a display querying them about certain numbers or species that seem unlikely based on the participant's location.

The data collected by CFW participants will initially be analyzed separately from the data collected by Project FeederWatchers to see how they compare. As the protocol for data collection is exactly the same for both projects, and as our initial impressions are that students collect data at least as accurately as backyard participants, we expect to find the data sets to be similar. In fact, the only real difference between these two projects is that Project FeederWatch assumes an ability to identify birds and relies on a simple instruction booklet, whereas CFW includes a full classroom curriculum, which begins by developing bird identification skills. In other words, FeederWatch will continue to be one big project, with two methods of implementation, and if desired, the CFW materials can be used by anyone interested, including individuals, informal groups such as Scouts and nature centers, and home schoolers. We imagine that data from both projects will frequently be combined for bird population analyses.

THE FUTURE OF PROJECT FEEDERWATCH

At the same time that development of CFW is being completed, the original Project FeederWatch will also be brought on-line in cooperation with the Cornell Theory Center. We feel that operating FeederWatch over the Internet will allow us to develop improved participant–scientist partnerships, and will solve the problems of timeliness in reporting results. The real-time data editing feature discussed earlier also should improve the reliability of data submission. We expect a test of new Web-based data submission and retrieval to begin early in 1997, and a listserve for participants and CLO staff to share ideas and findings to be developed around the same time. We will continue operation of Feeder-Watch with the original computer-scannable forms indefinitely, to accommodate participants not yet hooked to the Internet, but we also assume that within a few years' time, most participants will be able to participate on line if they choose.

THE RELATIONSHIP BETWEEN SCIENCE AND EDUCATION

We believe that one of the most exciting aspects of Project/Classroom Feeder-Watch, as well as other, large-scale projects that address questions of honest interest to science, is the interaction that results between scientists and educational program developers. The projects bring great benefits to both parties.

Scientists gain from such projects because they allow the answering of large-scale questions that cannot be addressed without large numbers of volunteer participants collecting information over wide geographic areas. Furthermore, when citizens participate in real science, their understanding and knowledge of the scientific process is increased, resulting in improved understanding of what scientists do. This creates a more hospitable culture in which scientists can work.

Program developers gain from such projects by plugging into ongoing, real-science issues that allow inquiry-based instruction to flourish in the classroom. It is easy to develop exciting programs for participants who can see the meaning of their data-collection efforts. Also, such programs allow a natural integration between many subjects including writing, math, geography, statistics, even music and art.

We are fortunate at the CLO to have strong, well developed programs in both science and education. Staff from both programs work closely together at all stages of project operation and development. This ensures that the science conveyed is correct and meaningful, and that participants understand what they are supposed to do, why they are doing it, what the data show, and how they have contributed to science. We believe that our citizen science model will be replicated most easily at other institutions having dual programs in research and science education, such as science and natural history museums.

CONCLUSION

The current emphasis on inquiry-based instruction, coupled with the idea that data generated by the public can benefit scientific research, has resulted in a powerful new model for science education. Scientists are discovering that students can act as eyes and ears across the country, gathering information on physical and biological characteristics such as vegetation, water quality, and wildlife populations, especially birds. Educators are learning that participants who gather such information improve their understanding of science content, acquire the skills of patient observation, imbibe the process of scientific investigation, and gain the satisfaction of furthering scientific knowledge. Finally, the evolution of Internet tools that can send, retrieve, and display large data sets, and that can connect scientists, teachers, and students for instantaneous communication, enhances the ability of these different groups to work together. CFW uses the Internet to bring together scientists, students, and their teachers to investigate real science questions in a manner unthinkable just one decade past.

ACKNOWLEDGMENTS

Project/Classroom FeederWatch depends on a dedicated staff at the Cornell Laboratory of Ornithology, including: Margaret Barker, Project Coordinator; Kenneth V. Rosenberg, Chief Scientist; Diane Tessaglia, Data Analyst/Computer Specialist; and Martha Fischer, Technical Associate. Also, Pixie Senesac served as Technical Associate during the pilot years. Judy Vesel, project director at TERC, is the author of the CFW curriculum, and Gloria Jacobs provides evaluation and technical support. Evaluation staff include Bruce Lewenstein, Laura Brown, and Rebecca Douglas. Project FeederWatch is a joint project of the Cornell Lab of Ornithology, the National Audubon Society, Bird Studies Canada, and the Canadian Nature Federation. Evaluation of Project FeederWatch and development of the Seed Preference Test were supported by NSF grant ESI-9155700, and development of CFW is supported by NSF grant ESI-9550541.

REFERENCES

Birders World, "Top Ten Feeder Birds," *Birder's World* 9(6)(1995):12.
Rick Bonney, "FeederWatch: From Baseball to Bird Seed," *FeederWatch News* 1(1988):1.
Rick Bonney, "Amateurs Welcome," *Living Bird* 10(1)(1991):36–37.
Rick Bonney, "Project FeederWatch," *Living Bird* 13(4)(1994):34–35.
Cornell Office of Communication Strategies, "Focus Group Analysis: Project FeederWatch/Seed Preference Test" (1994).
Cornell Office of Communication Strategies, "Project FeederWatch Survey: Report of Findings" (1995).

André A. Dhondt, "House Finch Disease Update," *Birdscope* 11(1)(1997):4.

André A. Dhondt and James D. Lowe, "Variation in Black-capped Chickadee Group Size," *Birdscope* 9(1)(1995):7–8.

Hezel Associates, "First-Year Evaluation of Project FeederWatch," 1201 East Fayette Street, Syracuse, NY 13210 (1992).

National Academy of Sciences, *National Science Education Standards* (Washington, DC: National Academy Press, 1996).

Kenneth V. Rosenberg, "Food For Thought," *Birdscope* 10(2)(1996):6.

Kenneth V. Rosenberg and Rick Bonney, "Birds Like Millet and Milo, Too!" *Birdscope* 8(4)(1994):8–9.

Diane L. Tessaglia and Kenneth V. Rosenberg, "Project FeederWatch: Annual Report 1993–94," *Birdscope* 8(4)(1994):1–6.

Diane L. Tessaglia and Kenneth V. Rosenberg, "Project FeederWatch: Annual Report 1995–96," *Birdscope* 10(4)(1996):1–5,8.

Jeffrey V. Wells, Kenneth V. Rosenberg, Diane L. Tessaglia, and André A. Dhondt, "Population Cycles in the Varied Thrush," *Canadian Journal of Zoology* 74(1996):2062–2069.

CHAPTER 4

The Science that Happens through Partnerships
Sun Photometer Atmospheric Network (SPAN)

Forrest M. Mims III

There is a long and honorable tradition of amateurs and students performing credible science by making and recording observations. Sometimes amateur scientists even make scientific discoveries. Consider Donald Parker, an anesthesiologist by day and amateur astronomer by night, who discovered very rare large-scale storms on Saturn in 1994. His work led to a major paper in *Science* in February 1996 (Sanchez-Lavega *et al.*, 1996)

When he was a fifth grade student in Los Angeles, Tommy Cragg became fascinated by astronomy during science class. Although he has always been an amateur astronomer, since 1944 Cragg has worked alongside professionals at Griffith Observatory, Mount Wilson, and Anglo-Australian Telescope facility in Australia. By the end of 1992, he had logged more than 132,000 visual magnitude estimates of variable stars. He has also continued an almost daily program of monitoring sunspots since 1944 (Levy, 1994).

Problems can occur when amateur scientists and students make discoveries before establishing a partnership with professional scientists. Scientists have sometimes been victimized by amateurs and students who make baseless claims or even create hoaxes. Thus, reasonable skepticism by the scientific community is appropriate. Sometimes, however, major opportunities are lost or delayed by the failure of scientists to check out the claims of a student or amateur or by

Forrest M. Mims III, 433 Twin Oak Road, Seguin, TX 78155; fmims@aol.com.

automatically assuming that nonspecialists are simply incapable of making original scientific discoveries.

Consider the classic case of Jerry MacDonald, who was working on his master's degree in sociology when he discovered the largest find of Permian-era fossil footprints on the planet. Even though he was ridiculed by several professional paleontologists at museums in New Mexico as a crank or worse, MacDonald persevered until he managed to convince representatives of the Carnegie Institution and the Smithsonian Institution to visit his site. They did—and they pronounced MacDonald's find as valid. As a result, some of his best finds now reside in Washington and Pittsburgh and not in New Mexico where they were discovered (MacDonald, 1992). In this case, the amateur's work was completely professional in every respect. I know, for I interviewed both him and scientists who validated his discoveries.

Francisco Garcia, an amateur astronomer in Spain, was sky watching one evening in March 1993 when he noticed an anomalously bright point of light in an arm of the famous spiral galaxy known as M81. Garcia's find turned out to be the brightest supernova discovered in the Northern Hemisphere sky since 1937. Alexi Filippenko of the University of California pronouned Garcia's find as the second most important discovery of a supernova of this century (*Sky & Telescope*, 1993). Other amateur astronomers made valuable contributions to Garcia's discovery. At midnight on March 27, M81 appeared normal in an image made by French amateur Claude Merlin. But 9 hours later it was clearly visible in an electronic image made by Bill Neely, an amateur astronomer in New Mexico (Pease, 1993).

Ida Kouasi, a citizen of Togo in Central Africa, completed my course in experimental science at the University of the Nations in Hawaii 2 years ago. Although Ida was a humanities major, the science course was required for graduation—and she was terrified of science. I know, she told me so many times. But after the course, this bright African woman asked if she could become part of the Sun Photometer Atmospheric Network. For more than a year Ida has been faithfully making daily measurements of ozone, column water vapor, UV-B, and aerosols using a hand-held instrument and mailing her observations. Ida's is one of only four ozone stations in Africa. One of the others is also a SPAN site. It so happens that Togo is a critical location for observing atmospheric aerosols, so Ida's observations have taken on a new importance.

Zeolites are minerals composed of highly porous aluminosilicates. Zeolites have valuable commercial uses as catalysts, especially in the cracking of petroleum into gasoline, where they increase the conversion efficiency by 10–20%. Six years ago a team of 18 amateur mineralogists spent 3 to 4 months of weekends searching for natural zeolites in a cold, wet hole near Washington's Columbia River. They discovered what came to be known as boggsite, a completely new zeolite with spectacular properties, and their finding was announced at a meeting of the Geological Society of America in November 1989 (Moffat, 1990).

Missionaries have a tradition of making scientific discoveries, and so do their spouses. Sarah Lyman was the wife of a missionary in Hawaii. A highly detailed diary she and her daughter-in-law kept between 1833 and 1917 represents the only continuous record of volcanic activity in Hawaii during the 19th century. Lyman's diary has allowed scientists to extend volcanic observations in Hawaii to well before the massive Kau earthquake of 1868. M. Wyss and his colleagues have even converted Lyman's highly detailed descriptions into the Modified Mercalli Intensity Scale. In 1992 they published their findings in "The Lyman Hawaiian Earthquake Diary, 1833–1917" (Wyss *et al.*, 1992).

Speaking of Hawaii's missionaries, the Reverend Sereno E. Bishop of Honolulu noticed a curious halo around the Sun after the 1883 eruption of Krakatau. He sent a report about his observation to *Nature*, which published it in 1884 (Bishop, 1884). Ever since, the phenomenon he observed has been called Bishop's ring.

Amateur astronomers have occasionally reported seeing what appeared to be white patches at the poles of Mercury. Of course this couldn't be right. Everyone knows that Mercury is much too close to the Sun to have water, much less ice caps, and everyone knows amateur observations can't possibly be better than those of the professionals, who once imagined canals on Mars. And when Mariner 10 flew by Mars in 1974 and 1975, it saw no ice caps. All this changed in 1991 when radar signals revealed what may be ice caps on Mercury. As Mercury is the only planet without a tilted axis and as Mariner 10 did not fly precisely over Mercury's poles, astronomers believe Mercury may indeed have ice. As Bob Berman wrote in *Discover* magazine in June 1994, "It's another case (like the strange radial spokes on the rings of Saturn) of keen-eyed amateurs seeing esoteric detail, in moments of steady air, that multimillion dollar observatories had completely missed" (Berman, 1994).

There are many amateurs among even the professionals in science. Remember the zeolite story from Washington? One participant was Rudi Tschernich, a postal worker. But another was Donald Howard, a physics professor at Portland State University. Howard might be a professional physicist, but when it comes to mineralogy, he and Tschernich are equals: they are both amateurs working outside their field.

Countless examples of this could be cited. Oxford mathematician and physicist Roger Penrose describes himself as a "dabbler" in various fields, a trait he acquired from his father who pursued psychology, philosophy, and mathematical puzzles. Former high school science teacher John DeLuisi went on to earn his Ph.D. in physics. He is now among the world's experts in the measurement of atmospheric aerosols, the ozone layer, and solar ultraviolet radiation. Down the hall from DeLuisi's office at NOAA in Boulder, Colorado, works Walter Komhyr, another physicist. Komhyr is world renowned among the ozone community for establishing and meticulously calibrating the Dobson instruments which now serve as standards for measuring the ozone layer and for inventing an accurate method for measuring ozone from a small weather balloon. Visit any laboratory, and you will find professional scientists gainfully employed in fields remote from their course of study during college days. Yet, like serious amateur scientists who may have never taken a course in science, they make valuable contributions in science.

By now I hope you're convinced that nonspecialist adults can still contribute to science and even make original scientific discoveries. So let's have a look at what students are doing.

Arthur Cook was not a professional scientist when he hypothesized that the small white stripe that appears at the top of the chrysalis of some monarch butterflies denotes a male. After carefully observing the sex of butterflies emerging from their chrysalises, Arthur confirmed his hypothesis. For this achievement, Arthur Cook's school received a national science award for his school. Arthur was 11 years old (Herberman, 1990).

Shortly after the launch of Sputnik 1 in 1957, Desmond King-Hele realized that the orbits of easily viewed artificial satellites could be predicted if he could simply receive enough reports from around the world. Amateur satellite watchers from around the world, many of them students, provided just the data King-Hele needed to predict the orbital decay of many satellites, He was even able to overturn previously untested theories about the density of the upper atmosphere and the effect of the sunspot cycle on the upper atmosphere using data provided by legions of enthusiastic volunteer observers (King-Hele, 1992).

Mark Morrissey and his wife Susan Postawko are meteorologists at the University of Oklahoma. Both of these Ph.D. scientists fully realize the vital role that can be played by students in collecting atmospheric data. The Schools of the Pacific Rainfall and Climate Experiment (SPaRCE) they founded now collects rainfall and other data from schools on more than 30 islands in the Pacific Ocean. Data from this program have already been published in scientific papers.

One of the most fertile domains of amateur science is ornithology. People may sometimes make fun of amateur bird watchers, but serious observers of birds have made major contributions to ornithology. Cornell University realizes the valuable role played by amateurs, for in 1993 Cornell ornithologists had recruited more than 8000 volunteers in North America to survey bird feeders, make recordings of birdsongs for the Library of Natural Sounds, count migrating raptors at hawk-watch sites, and take part in the annual Christmas Bird Count (Holden, 1993). Do these volunteers contribute to real science? In 1992, 2265 volunteers in 61 states and Canadian provinces participated in a Cornell pilot study to determine if ground-feeding birds like red milo and black-oil sunflower seeds, which are commonly used in commercial bird feeds. A project like this might seem mundane, but it is important to programs that are attempting to attract birds to specific areas for study and observation.

I could describe many other projects in which amateurs and students are making important contributions to science. There's the Audubon Society Christmas Bird Count, which has supplied invaluable data used in many papers by ornithologists. There's the Massachusetts Acid Rain Project, carefully managed by Paul Godfrey. There's the University of Kansas Monarch Watch, in which my youngest daughter and I have participated and about which we will learn much more at this meeting. There's EARTHWATCH, which annually sends hundreds of volunteers to help scientists conduct research all over the world. And there's the Sun Photometer Atmospheric Network, a global network of volunteers who use a sophisticated, hand-held instrument, developed with resources from a 1993 Rolex Award for Enterprise, to measure the ozone layer, solar ultraviolet-B, and the aerosol optical thickness of the atmosphere.

DOING SCIENCE

Certainly we all expect any science conducted by students to be legitimate science, whether or not it is publishable. There must be no violation of animal and human protocols, no fudging of data, no cheating, no fake experiments, and no failure to acknowledge the work of others. And there should be no exploitation of students by scientists. Unfortunately, that may happen to them in graduate school anyway, and we need not speed up the inevitable.

There should also be a better way of explaining the scientific method—whatever that is—to students. Several years ago I described the scientific method in a Radio Shack comic book distributed to 200,000 students (Mims, 1993):

1. Select an idea.
2. Decide on the purpose of your project—what specific problem do you want to solve or what question do you want to answer?
3. Form a hypothesis about your project—a brief statement about the project experiments you will do and a very general guess about the possible results.
4. Plan your experiments or observations.
5. Collect the necessary supplies.
6. Begin your experiments or observations.
7. Carefully record your results and make tables and graphs.
8. Study your results. What happened? Why? How?

All of these steps are completely valid, but one is optional: "3. Form a hypothesis about your project." Sometimes I form a hypothesis about a particular research project. For example, when I measured the ozone layer during two solar eclipses, I assumed that it would be possible to detect waves in the ozone caused by the cooling of the moon's shadow. It was possible, and my son Eric and I published a paper about our measurements in a peer-reviewed journal (Mims and Mims, 1993).

But usually I do not formulate a hypothesis. And I rarely try to anticipate a result! Last summer NASA sent me to Brazil to measure the ozone layer, the Sun's UV-B, and the optical density of smoke during the burning season. I was quite surprised to find an 80% reduction of UV-B caused by smoke from biomass burning and a 40% reduction in visible sunlight (Mims, 1995, 1996a).

Although this research was begun with no hypothesis, it has led to the hypothesis that thick smoke from biomass burning and severe air pollution might enhance the population of disease-transmitting mosquitoes and pathogenic bacteria on exposed surfaces and suspended in air and water (Mims, 1996a). This, in turn, led to very simple and inexpensive ways to determine the spectral response of mosquito larvae and the preference of adult females for dark sites to deposit their eggs (Mims, 1996b,c).

Most of my research is not preceded by a hypothesis. In 1994 I placed 14 UV-B data loggers around the Big Island of Hawaii to study the effect of altitude on UV-B. My interest in altitude effects disappeared when some of the data loggers showed very large increases in UV-B caused by sunlight scattered from the sides of cumulus clouds. This serendipitous finding, which led to a paper in *Nature*, was completely unexpected (Mims and Frederick, 1994).

For reasons such as these, the almost universal requirement that students formulate a hypothesis and anticipated results should be relaxed. I have judged many science fairs and interviewed lots of discouraged students who felt they had failed when their hypothesis could not be validated. Based on my own experience, the best discoveries come from observations and serendipity, not a planned hypothesis. The hypothesis should not be eliminated, just made optional.

One more thing about science deserves much attention: It isn't necessary to spend lots of money to do publishable science and even make discoveries. A few years ago I was often told that my work would be taken more seriously if the instruments I develop were more expensive! After results of measurements with these instruments began to appear in peer-reviewed journals, my inexpensive methods were taken much more seriously by various scientists, some of whom have since assigned me to conduct various kinds of solar and atmospheric measurements in Brazil and around the United States.

SOME STRATEGIES FOR IMPLEMENTING SSPs

So if you agree with me that students can indeed learn science by performing science, what's next?

First, the current approach to SSPs is to toss assorted projects out to teachers who have little time and even less ambition to depart from state-mandated programs. This method will always limit participation in SSPs. Instead, we should find ways to seamlessly integrate appropriate SSPs into the overall science curriculum so that they become a traditional part of one's science education. Cornell's FeederWatch program and the University of Kansas's Monarch Watch are excellent examples of the kinds of basic SSPs that have been extensively tested and that do work.

Next, SSPs must include follow-up studies that teach fundamentally important scientific principles. Counting birds and tagging monarchs is fine, but serious learning takes place when students learn how to apply such tools as simple statistics, mapping, and climatology to the collective findings of all of the participants in a particular SSP.

We should establish a goal that findings of student participants in SSPs will be published on the Internet's World Wide Web and incorporated in papers published in peer-reviewed journals of science. This will certainly increase the interest of students. And it will disarm or at least neutralize any scientists who are skeptical that students can conduct real science.

We should also publicize the fact that results from existing SSPs have already appeared in peer-reviewed astronomy, ornithology and meteorology journals. For example, recent papers in meteorology journals have used data from the 32-school Western Pacific network operated by SPaRCE.

Some scientists will oppose SSPs. Of course skeptics serve a vital function in evaluating SSPs as they will force us to be careful. Therefore, every SSP should simply assume that it may one day be the target of a critical scientist and do its best to maintain the highest standards.

Students understand these kinds of things. They are at least as good at exploiting rivalries as adults, and they know exactly what a scientist is thinking who learns that a couple of students are using $500 hand-held instruments to make ozone measurements at least as accurate as those he or she makes with a $100,000 instrument. Students also know what a scientist thinks when he learns that they collected data for 90 consecutive days, but his data have mysterious 2-day gaps at 7-day intervals plus occasional 7 to 14 day gaps. Could weekends and vacations be getting in the way of science? In short, students understand the phrase rarely heard but often practiced in science: protecting one's turf.

But students may not understand that a few professional scientists may view them as children with no credentials who are attempting to use toy instruments to achieve what he or she spent long years learning to do. I suggest that we bypass or avoid interrupting the work of these scientists and press on without them. In the final analysis, even they can't ridicule SSPs that result in peer-reviewed publications.

So far I've addressed only the students and scientists of the SSP concept. But teachers can, if they wish, form an equally important term in the SSP equation. This is simply one more good reason to reform and reinvigorate the science teaching establishment. It's in their own best interest.

There are many ways to reform the science teaching establishment, but I will mention just two:

1. Simplify the requirements for teacher certification to entice trained professional scientists and retired scientists and engineers into the profession. The current situation is such that a college science instructor who writes science textbooks and loves to teach and work with students isn't qualified to teach high school science.
2. As a condition of certification, science teachers should be required to complete and publish on the Web or in a journal a quality science project. For the current generation of science teachers, many of whom have never actually conducted a science project, require a summer workshop in which each teacher completes a quality science project.

Of course there will be objections to upgrading the current crop of science teachers. Even some of the staff at colleges that train teachers may object, especially those who have not bothered to set foot in a secondary school classroom for the past decade or so. They might be especially concerned about my dream that every science educator—at all levels—be involved in some kind of ongoing science study or project. A powerful incentive for those who object is to remind them of the home school explosion, which now includes an estimated 1 million students, and the imminent arrival of the CCD and Web-based virtual classroom.

I realize that both SSPs and reforming science teaching cost money. The new era of communications technology offers spectacular capabilities, but going on line is expensive for the many schools that have primitive or even nonexistent computer capabilities. Until the Internet is universal, we must realize most schools are not yet on the Net. Therefore, it is important to include in the mix low-budget SSPs that do not require Internet access. Schools with little money can follow the example of Professor John H. Bedenbaugh of the University of Southern Mississippi, who has developed very simple and inexpensive chemistry demonstrations and experiments for poor school districts in his state (Bedenbaugh, 1991).

It's also urgent to develop a strategy to convince parents that the future of their children will be enhanced more by better science education than by new athletic uniforms.

Finally, SSPs must in fact contribute to a student's overall knowledge base and scholastic achievement. Otherwise students will be victims of just another educational fad, which is what SSPs will be considered if they fail to enhance a student's level of academic achievement.

Related to this is the community's overall perspective of SSPs. If SSPs are designed or used to implement various political agendas, for example, they will simply antagonize many parents who already feel that public schools offer a biased curriculum. For example, I can assure you from personal experience that there is great pressure to transform certain kinds of environmental research from objective to agenda-driven science. I have been asked on several occasions by representatives of groups from various sides of the political spectrum to make statements or express conclusions that simply cannot be supported by the atmospheric measurements I have made. If SSPs avoid political agendas and simply perform objective, credible science that is published in peer-reviewed journals, parents will gain a new sense of pride in their children and schools.

AN SSP WISH LIST

Finally, allow me to close with a dream. Here is my wish list of SSPs:

1. Every school is a weather station.
2. Many schools are seismic stations.
3. Every school measures background ionizing radiation.
4. Every school measures solar UV-B.
5. Every school measures haze and the optical thickness of the atmosphere using a Sun photometer such as the Global Lab VHS-1 (Mims, 1996d).
6. Every school measures polarization of sky.
7. Every school measures pH of precipitation.
8. Every school annually surveys the local population of birds, insects, and plants.
9. Every school measures bacteria colony forming units in the outdoor air (Mims, 1996d).
10. Every student participates in project E.T. by measuring the Sun's extraterrestrial constant using the Langley method (Mims, 1992).
11. Every school participates in the Tom Jefferson Project (Mims, 1992).
12. Every science teacher has completed a quality science project.

Students who actively participate in projects like these under the guidance of science teachers who have actually done science will be far better prepared for the world ahead no matter what their chosen field.

REFERENCES

John H. Bedenbaugh, "How to Measure the Oxygen Content of Air," *Science Probe* 1(1991):75–78, 120.

Bob Berman, "Mercury's Rising," *Discover* 15(1994):55.

Sereno E. Bishop, "The Remarkable Sunsets," *Nature* 29(1884):259–260.

Ethan Herberman, *The Great Butterfly Hunt* (New York: Simon & Schuster Books for Young Readers, 1990), p. 27.

Constance Holden, "Amateurs Gather Avian Data," *Science* 262(1993):1976.

Desmond King-Hele, *A Tapestry of Orbits* (London: Cambridge University Press, 1992).

David H. Levy, "Tommy Cragg: Watcher of the Sun and Stars," *Sky & Telescope* 87(1994):101

Jerry MacDonald, "Footprints from the Dawn of Time," *Science Probe* 2(1992):32–47.

Forrest M. Mims III, "Surveying Your Environment," *Science Probe* 2(1992):3.

Forrest M. Mims III, "Science Fair: Ticket to Your Future," Radio Shack (1993).

Forrest M. Mims III, "Smoke and Rainforests," *Science* 270(1995):5243.

Forrest M. Mims III, "Significant Reduction in UV-B Caused by Smoke from Biomass Burning in Brazil," *Photochemistry and Photobiology* 64(1996a):123–125.

Forrest M. Mims III, "UV Radiation and Field Experiments," *BioScience* 46(1996b):564–565, 1996.

Forrest M. Mims III, "Biological Effects of Diminished UV and Visible Sunlight Caused by Severe Air Pollution," Proceedings of the International Radiation Symposium, Session 8, Fairbanks, Alaska, 19–24 August (1996c).

Forrest M. Mims III, "TERC Global Lab Visible Haze Sensor (VHS-1)" (1996d).

Forrest M. Mims III and John E. Frederick, "Cumulus Clouds and UV-B," *Nature* 371(1994):291.

F. M. Mims III and E. R. Mims, "Fluctuations in Column Ozone During the Total Solar Eclipse of July 11, 1991," *Geophysical Research Letters* 20(5)(1993):367–370.

Anne Moffat, "Rocking on the Banks of the Columbia River," *Science* 247(1990):1412.

Roland Pease, "Supernova Brightens the Horizon," *Science* 362(1993):585.

A. Sanchez-Lavega, J. Lecacheux, J. M. Gomez, F. Colas, P. Laques, K. Noll, D. Gilmore, I. Miyazaki, and D. Parker, "Large-Scale Storms in Saturn's Atmosphere During 1994," *Science* 271(1996):631–634.

"Supernova in *M81*," *Sky & Telescope* 85(1993):8.

M. Wyss, R. Y. Koyanagi, and D. C. Cox, "The Lyman Hawaiian Earthquake Diary, 1833–1917," *U. S. Geological Survey Bulletin* 2027(1992).

CHAPTER 5

The Aquanaut Program

Ivar G. Babb, Peter M. Scheifele, and Domenic Tedeschi

INTRODUCTION

The Aquanaut Program (AP) began in 1988 as an educational initiative of the National Oceanic and Atmospheric Administration's (NOAA) National Undersea Research Center at The University of Connecticut at Avery Point (NURC-UCAP). At its inception, the AP was a purely experiential program with students and teachers spending one day on board a ship observing scientists and diving in an occupied submersible. In response to feedback of participating teachers the AP evolved into a more rigorous, yearlong project that includes introduction to a research topic, interactions with a research mentor, hands-on field research during the summer months, and subsequent data analysis and presentation during the fall semester. The AP is open to all secondary schools in the Northeast Region served by NURC-UCAP and typically supports about 100 students and 20 teachers per year. During its history, the AP has enrolled 885 students and 128 teachers from schools throughout the regions served by NURC-UCAP. This includes 28 students from high schools in Michigan who have been involved with a similar program in the Great Lakes and 19 students from Israel.

Program Goals

AP's goals are:

1. To enhance science literacy and encourage students to pursue the study of science

Ivar G. Babb and Peter M. Scheifele, National Undersea Research Center, University of Connecticut at Avery Point, Groton, CT 06340; scheifl@unonvm.uconn.edu.   Domenic Tedeschi, Norwich Free Academy, Norwich, CT 06360.

2. To provide students and teachers an inquiry-based opportunity to experience the research process under the guidance of a mentor scientist
3. To provide high school teachers an opportunity to enhance their science teaching skills by participating in a research project utilizing modern technologies
4. To develop a sense of stewardship for the marine environment in today's students

Program Approach

The AP relies on three characteristics that contribute to its success: (1) the stimulating nature of *advanced technologies* used in modern oceanographic research [e.g., remotely operated vehicles (ROVs), occupied submersibles, sonar systems], (2) strong *scientist/teacher/studentlinkages*, and (3) relevant environmental *researchtopics*. These attributes combine to immerse teachers and students in the culture of research, allowing them to experience the entire research process. The following sections describe each of the major facets of the AP.

ADVANCED TECHNOLOGIES

The oceans are formidable and complex. They represent a hostile environment for humans, characterized by the incredible pressure of hundreds of feet of water, near freezing water temperatures, little or no light and strong currents. The study of the oceans requires many advanced technologies to withstand these rigors. To provide human presence throughout the water column and on the ocean floor requires the use of specialized "*in situ*" technologies. Terrestrial environmental education has benefited from the ease with which forests, deserts, shorelines, and lakes and streams can be accessed. A major problem with marine studies is that the environment is not readily accessible to the average teacher or student in real time. Human presence can only be provided directly through the use of occupied submersibles and scuba diving or indirectly through the use of ROVs and sonar equipment (representing mechanized eyes, hands, and ears). The AP was founded by the NURC-UCAP, which is one of six centers established by NOAA's National Undersea Research Program (NURP) to provide these technologies to research scientists based on a competitive proposal review process. NURC-UCAP strongly supported the development of the AP and currently supports the program based on a similar competitive educational program review process established by the NURP.

ROVs as Educational Tools

The AP historically relied on occupied submersibles to provide an undersea experience for high school students. These systems, although exciting and appropriate for many sampling tasks, are expensive and limit the number of participants to two observers and a pilot. Presently, ROVs are used as the primary, hands-on data collection tool to carry out the educational and research goals of the AP. Several characteristics of ROVs provide distinct advantages over other oceanographic tools in meeting these goals (Babb and Scheifele, 1992).

ROVs were developed to provide an alternative to occupied submersibles in commercial offshore oil and gas operations. They are controlled by a tether or cable that provides power and command controls to the vehicle, while providing a conduit for the return of video and data signals (see Fig. 1). An operator controls or "flies" the ROV through a command console located on board the vessel. This controller often resembles the joysticks found on computers and arcade games, which promotes rapid acclimatization by students. This controller directs the operation of thruster motors with propellers to control the movement of the ROV. The video signal can be broadcast on multiple monitors and recorded on multiple videotape decks for subsequent viewing and analysis. The ROV can be maintained underwater for greater periods of time than a diver or submersible, and is limited only by the ability of the vessel to maintain station over the ROV and the stamina of the operator(s) and viewers. In addition, the cable provides wide bandwidth for the transmission of additional data such as audio signals, conductivity, temperature, depth, dissolved oxygen, pH, and other variables. Several of the ROVs operated by NURC-UCAP are lightweight and portable, making them extremely useful from a variety of vessels that operate in local waters.

The ROV is an excellent hands-on educational tool as it provides real-time control of the vehicle by a teacher or student, including control of a simple robotic manipulator (mechanical arm), video and still photographic cameras. Control of the ROV is not a simulated experience, but a real-world activity that provides teachers and students the thrill of discovering underwater creatures and allows them to observe and follow them using the maneuvering capabilities of the ROV. It also provides them with an understanding of the limitations of the technology and the frustration of working in local underwater environments, often characterized by high currents and low visibility. Unlike an occupied submersible that allows one or two observers an isolated opportunity to see the ocean environment, the ROV allows many teachers and students to view the ocean floor in the company of their teacher and research scientists. This promotes interactive dialogue among all participants to discuss the real time images observed, including ocean floor habitats, species identification and ecological role, unique behaviors, physical features, and events.

The ROV provides students and teachers with the ability to view and collect samples from local marine environments in the study of a current research topic, which promotes environmental stewardship. This ability to see and sample ocean floor environments also fosters the "environmental sensitivity" variable defined in the environmental behavior model developed by

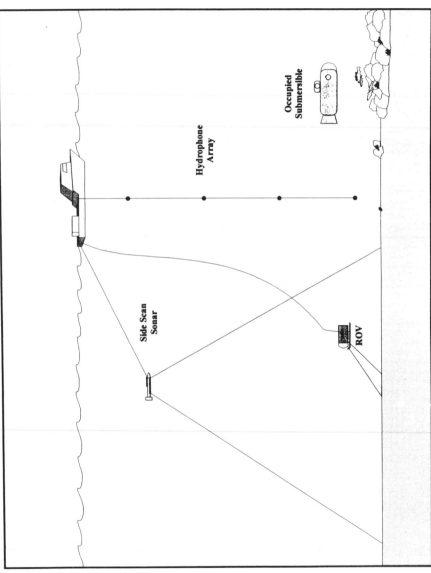

Figure 1. One of the key elements in the success of the Aquanaut Program is students being provided the opportunity for hands-on operation of multiple advanced technologies.

Hungerford and Volk (1990) by allowing the students to experience either pristine or degraded habitats. The research themes defined by the participating scientists combine with the visual capability of the ROV to provide a practical understanding of the ecological concepts structuring these local waters. In addition, the fact that the students are studying in their own backyards promotes the "ownership variables" described in Hungerford and Volk's model. The "personal investment" variable enforces the relevance of the research that the students are involved with, which, in turn, promotes the sense of empowerment that the students can and are making a difference in their contributions to the research theme (Fig. 2).

The ROV also is a powerful research tool that provides research scientists working with the AP a new perspective on their research. In some cases the ROV aids in ongoing research projects by adding a visual perspective that was lacking previously. In other cases the research mentor can use the visual information provided by the ROV as the first order observations (i.e., natural history) needed to initiate a new research project. The ROV is ideally suited to data collection by the students involved in two of the research themes supported by the AP. For the bioacoustic study the ROV provides a visual validation of the source of biological sounds recorded using hydrophones mounted on the ROV. The ROV rests on the ocean floor in a quiet observational mode during acoustic monitoring to provide simultaneous real-time visual information in tandem with the sounds being recorded. For the Stellwagen Bank study, the ROV provides a tool to record the distribution and abundance of the dominant seafloor creatures using both video recordings and still photography.

Other Technologies Used by the AP

In addition to being a difficult environment to work in, the oceans are also very complex, requiring that the study of oceanography is inherently multidisciplinary, involving concepts from biology, chemistry, physical science, mathematics, computer science, and earth science. Each of these subject areas provides opportunities to apply concepts learned in the classroom. The collection of data to support the study of these disciplines in the oceans requires the use of many technologies. The following are additional exciting technologies utilized by the AP in support of its research topics described below.

Multiple types of sonar gear have been used extensively to support recent AP research themes. Advanced active/passive sonar suites have been used in the Ambient Noise theme. These sensitive devices are deployed over the side of the vessel as well as from small craft launched from the mother vessel and are capable of recording sounds from miles away. The acoustic data are analyzed by a spectrum analyzer and computers using bioacoustic software, operated by student Aquanauts to measure background noise levels in the oceans as well as the sounds produced by whales, dolphins, and other marine animals. These data give students and teachers new insights into a discipline that is often overlooked in the study of the oceans, and one that requires extensive application of physics and mathematics.

The active portion of the sonar system enables the students and teachers to ensonify a section of the water column with high-frequency signals for the purpose of detecting the presence of small planktonic organisms and for conducting research on sound absorption in seawater. The passive (listening) portion of the sonar suite is used to record and monitor noise levels and noises made by fish and marine mammals for further analysis in the laboratory. Students have integrated the video capabilities of the ROV with the sonar in an effort to engineer the simultaneous visual and acoustic monitoring of benthic organisms.

Side scan sonar is a towed, torpedo-looking device (Fig. 1) that has acoustic transducers mounted on its sides that produce an acoustic pulse that travels to the ocean floor. This pulse is then reflected from the bottom, received by transducers mounted on the sonar, and recorded as an image on a video screen and as a paper record. The nature of the image depends on the strength of the returned acoustic signal, which in turn, depends on the hardness of the seafloor. Therefore, hard rock bottoms reflect more of the acoustic signal which is recorded as a darker image, whereas soft bottoms absorb a lot of the acoustic signal which is recorded as a lighter image. Oceanographers and geologists use this tool to provide a larger view of the sea floor, similar to the imagery that satellites provide of the earth's surface. Side scan sonar has been used recently by the Aquanauts to target certain bottom types for subsequent assay using the ROV.

More traditional oceanographic tools are also used by the AP to provide additional data to support the research themes. Students deploy, recover, download, and analyze data from a "CTD" used to measure the conductivity (which is reported as salinity), temperature, and depth. This information is critical for acoustic studies and is displayed as a profile with increasing depth. Opening and closing plankton nets are also deployed, recovered, and sampled by students to collect plankton samples from discrete water strata, which provides information for the acoustic research and monitoring themes. Students analyze the plankton samples using a stereoscope with attached video camera.

Each of these advanced technologies, the ROV, sonars, CTD, and video records of plankton, provides interesting, relevant data resources. One of the goals of the AP is to provide participating teachers new, updated, and advanced teaching media that can be brought back and employed in any classroom. This allows the benefits of the AP to be widely shared with a broad audience of teachers and students. In addition, this outreach is being further enhanced as each of the research themes data and findings are input to the AquaNet World Wide Web (WWW) home pages (see below).

LINKING EDUCATION AND RESEARCH

Teacher Professional Development

The AP was developed to bring the scientific and academic communities together. The AP is a concentrated, real-world program that immerses teachers and students into a true cognitive apprenticeship, working with a research practitioner. Teachers are involved in a professional development program that seeks to team teachers with a practicing research scientist to address his or her research specialty. Participants are involved with authentic activities that occur in the field and bring back the excitement of discovery as well as new skills to the classroom. The teachers are immersed in the culture of the research community, and the rigors of scientific methodologies used to address critical environmental issues.

The research scientists/mentors meet with the teachers early in the spring semester to assign and present the research topics, the background and significance of each project, the primary literature, and the methodologies and analyses to be used. The summer field program is held on 5 days in July or August. The teachers are involved in all facets of the research process including: (1) literature/WWW search, (2) research methodology development, including statistical design and methods, (3) data collection and analysis, (4) data presentation and means to "publish" data on the WWW, and (5) developing means to share data sets and integrate data into classroom activities. The actual summer schedule involves 1 or 2 days of preparation, 2 days on the water, and 1 or 2 days back in the laboratory analyzing data, sharing and developing concepts and materials to integrate into the classroom. At a mutually agreed on time in the fall semester all teachers attend a workshop. Past years have seen teachers demonstrating handmade transducers, computer models, and water column/biological monitoring procedures. This follow-up session provides an opportunity to share data collection, analysis, presentation, and potential classroom applications of the material. Each group provides a written report of its research and the educational materials or concepts developed from the project. This workshop also provides a debrief/critique opportunity for the teachers to provide input to the AP management staff.

Research Scientists' Opportunities

The research scientist provides his or her expertise and research topic to be implemented through the AP. In this fashion, the scientist is both a mentor and a role model for the students involved. The interaction has also been a two-way street. From the research scientists' perspective, the AP provides a unique opportunity to apply the technologies available through NURC-UCAP to their research interests. In the past, this has provided scientists with additional data to add to their existing projects or provided a first-time look at new areas as a pilot study. In some cases graduate students have assisted the research scientist, which provides the Aquanauts with additional role models as well as an opportunity for the graduate students to teach the research topics being addressed by their advisor. Finally, the direct involvement of the research scientist in the data collection and analysis provides strict quality control such that the results of AP research have provided useful information to many environmental management agencies (e.g., NOAA's National Marine Fisheries and Stellwagen Bank National Marine Sanctuary; the Environmental Protection Agency; and the U.S. Army Corps of Engineers).

The instruction on, and use of, the technological tools for executing the research project provides useful material to take back to the classroom to incorporate into existing science curricula. Teachers explore means to post raw data on the WWW in near real time and develop ways that these data can be shared and compared with other data sets as part of their integration into the classroom.

AP RESEARCH

Research Sites

Since its inception the AP has focused its research activities at two sites in the Gulf of Maine (Fig. 3): Stellwagen Bank (70° 27′ W, 42° 26′ N) and Pigeon Hill (70° 14′ W, 42° 46′ N) on Jeffreys Ledge. These sites were chosen for several reasons: (1) proximity to shore and port facilities allows day trips and helps ensure that the sea state is generally favorable, (2) the diversity of flora and fauna promotes a range of research opportunities for student projects, and (3) these areas have a long history of research and natural resource utilization. Recently, Stellwagen Bank has been named a National Marine Sanctuary, and as such, the AP is providing all of the data collected to the Sanctuary Manager to promote research and educational programs. Stellwagen Bank has been, and will continue to be, the site of other agency research efforts that have provided the AP with opportunities for principal investigator (PI) interaction, including the Environmental Protection Agency, the International Wildlife Coalition, and the U.S. Geological Survey.

All of the AP's research and educational activities are being maintained in a spatial database utilizing the ARC/INFO Geographic Information System (GIS). This computer-based system provides a powerful tool for the analysis and presentation of data. Specialized maps can be developed that present individualized data sets according to research theme. For example, maps have been made that show the relationship of dives from one year to another, the different levels of noise in the ocean from one area to another, and the spatial relationship of material dumped in the ocean to the actual designated dump site. This educa-

Figure 2. The Aquanaut Program Model short-cuts the typical flow of information from the research enterprise to the classroom through reliance on research mentors and a sense of ownership in the research being conducted.

tional tool is being used in each of the research themes and is encouraged as another advanced technological tool in the development of teaching resources and data/information management that can be used by the teachers. The GIS is also used to develop maps that are published on the AP home pages.

AP Phase I Research Themes

Since 1990 the Phase I portion of the AP has built a core of four research themes that address long-term environmental issues and that build on an expanding AP database. These research topics represent one of the major strengths of the AP, for each theme addresses a current environmental concern, they are relevant, and develop concern (i.e., a sense of ownership) from participants. These research themes require a multidisciplinary approach and therefore provide rich opportunities for integration within the general science concepts taught in the classroom. The real-world, problem solving nature of the research projects quickly answers students' questions as to why they need to learn mathematics or science concepts. The following discussion provides synopses of the core AP Phase I research projects.

The Pigeon Hill/Stellwagen Bank Monitoring Program. This is the longest running core theme for the AP, having been in existence for 7 years. It illustrates the value, rationale, and methodology of a long-term environmental monitoring program. Among the research and educational topics addressed through this study are the concepts of spatial and temporal variability, sample and population size, and statistical analyses. The research involves the collection of an "indicator" species, the sea scallop (*Placopecten magellanicus*) from the Pigeon Hill area on Jeffreys Ledge in the Gulf of Maine (Fig. 3) using the robotic arms of occupied submersibles or ROVs. Teachers and students interact with research personnel from NOAA's Milford Fisheries Center to dissect the animals and collect tissues for heavy metal contaminant analyses (e.g., Cu, Cd, Hg, Ag, Pb) at the fisheries laboratory. In 1996, the monitoring program expanded to include measurement of heavy metals in zooplankton species at several locations. In 1991 and 1992, Dr. John B. Pearce of NOAA's Northeast Fisheries Center was the mentor for this program; since 1993 Dr. Fred Thurberg of NOAA's Milford, CT Fisheries Center, Mr. George Sennefeld of NOAA's Milford, CT Fisheries Center, and Mr. Dom Tedeschi, AP Research Coordinator, have served as research mentors.

Stellwagen Bank Ambient Noise and Bioacoustic Impacts on Cetaceans. This second core theme for the AP began in 1992. It constitutes an introduction to physical and acoustic oceanography as they relate to marine biology and monitoring of the marine environment. Teachers and students operate an active/passive sonar to produce ambient noise curves for the area that are used

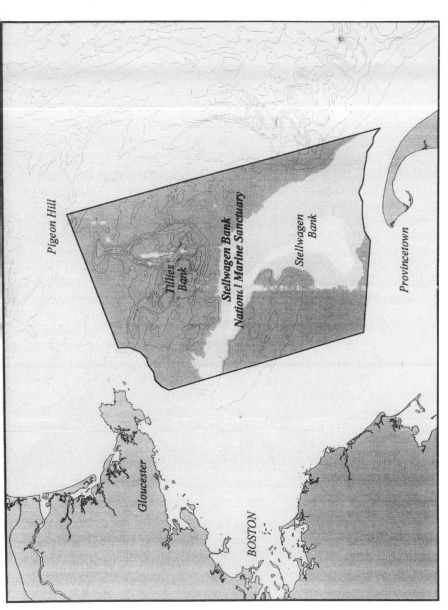

Figure 3. Map of the Gulf of Maine showing the Aquanaut Program Research Sites on Stellwagen Bank and Pigeon Hill along with the boundary of the Stellwagen Bank National Marine Sanctuary.

to assess the noise levels produced by specific activities such as shipping, commercial fishing, and recreational boating. The acoustic environment is completely characterized with the derivation of sound velocity curves based on on-board water chemistry and CTD (conductivity, temperature, depth) data. The overall intent of this acoustic characterization is to test two hypotheses (1) that with increasing vessel traffic in the area, the level of ambient noise is increasing and (2) that this increased level may be related to the recent decline of the cetacean population on Stellwagen Bank. The PI is Peter M. Scheifele, Director of Marine Education Programs/Bioacoustic Research at NURC-UCAP.

Stellwagen Bank National Marine Sanctuary (SBNMS)—Habitat Ecology of Fishes. In this theme, teachers and students operate an ROV to conduct video transects which are subsequently analyzed to determine how habitat structure relates to the distribution of fishes. Side scan sonar records are used to select survey sites. Participants also conduct statistical analyses to quantify the relationship between habitat type (e.g., mud versus sand versus rocks) and the animals that live in these different bottom types. This research is part of a larger NURC-UCAP initiative to assist the SBNMS natural resource assessment. In addition, the video and photographic data will be provided to the National Marine Sanctuaries Program as part of their management/education database for Stellwagen Bank and will be available for teacher use. The PI is Peter Auster, NURC-UCAP Science Director.

Hard Rock Dump Site Assessment. This study combines side scan sonar and the ROV. First, side scan sonar surveys are conducted and the records are analyzed to locate specific targets. Second, the ROV is used to conduct video transects to survey these targets. The area of interest is an area of newly dumped (1992) hard rock material adjacent to the northwest boundary of SBNMS. Students are involved with comparing the community developing on the hard rock area which now serves as an artificial reef with the adjacent, original substrate. They are also comparing the new reef fish community with existing, similar hard rock sites on nearby Tillies Bank. This area serves as a natural laboratory to address ecological themes such as succession, competition, and predation as well as providing resource managers valuable information on the impacts of this dumping activity. The research mentor for this project is Mr. Darryl Keith of the Narragansett Branch, EPA. He has been assisted by Mr. Ivar Babb, NURC-UCAP Director, and Peter Auster, NURC-UCAP Science Director.

AP Phase II Research Themes

The Phase II portion of the AP involves students who have completed the Phase I portion of the program and feel that they will pursue college-level studies and perhaps careers in mathematics, science, or engineering. The themes selected for Phase II typically involve more independent study by the students, working as individuals or small groups.

AquaNet: AP WWW Pages. The WWW represents one of the most exciting means for information exchange ever developed. The information superhighway is used extensively by researchers, educators, and commercial industries. As technology improves, greater transmission bandwidth is allowing rapid dissemination not only of text, but also of multimedia, including photographs, images, animations, audio, and digital video. The AP's research topics typically produce a great deal of multiple media data that can be directly applied as source material for the development of WWW pages. Furthermore, one of the criticisms of the AP is that, although it provides a rich, hands-on educational experience, the number of participants is low. The development of interactive WWW pages that provide data and materials on line is an attempt to disseminate the information, if not the excitement, of the program.

This Phase II AP project consists of the design, testing, and evaluation of AP WWW pages. The overall goals of this project are to present background material for direct use by AP teachers and students, to provide AP materials and data to schools nationwide to enhance or complement their existing environmental education activities, and to promote the AP. The objectives of the project are: (1) to have students design and develop an AP WWW site, (2) to learn to search the Internet for current marine data relevant to the AP research themes and integrate links to these sites to the AP page, (3) to add data collected from previous years to the AP pages, and (4) to collect and add new data from the AP research projects to the AP Web pages. Currently, the AquaNet home page (URL: http://ap.nurc.uconn.edu/www/aquanet.html), the Gulf of Maine page (URL: http://ap.nurc.uconn.edu/www/gomintr.html), and the AP Monitoring Program page (URL: http://ap.nurc.uconn.edu/monitor) are under construction.

Teachers will have a page for suggesting means to analyze, graph and present the collected data as well as related teaching activities. The teachers' page will also suggest activities to test students' understanding of the material.

The Gulf of Maine Page. This page provides a broad overview of the Gulf of Maine (GOM). For the WWW page to maximize interest and participation, students are developing numerous maps, photographic images, sounds and videos of the GOM. In addition, as there are many fine pages dedicated to the GOM, students have been surfing the WWW to locate and bookmark sites that they feel will be useful for other students interested in learning about this region. Among the features of the GOM that have been found to date are pages dealing with: (1) the physical boundaries of GOM, (2) bottom topography, (3) geological formations of the GOM, and (4) water flow patterns.

Virtual Tour of the Gulf of Maine. The most exciting and original aspect of this page is the Virtual Underwater Tour of the GOM. This tour is based on an image map of the GOM, highlighting two major areas: Stellwagen Bank and George's Bank. During the past year, students have been involved in a multistep process to develop this tour. First, they used a Geographic Information System to identify NURP-sponsored science missions conducted at these two areas. Second, they reviewed videotapes of the submersible or ROV dives from these missions and identified typical and unique images. Third, they used an inexpensive image grabbing device (Snappy) save still images from the videotapes. Fourth, the images were linked to an image map of the GOM that has points indicating the location of the mission. Clicking on a point brings up small snapshots of the images that can be expanded to full page images by clicking on them. In addition, background information on the mission (e.g., scientists name, e-mail address, mission objectives, technology used) will be provided as a link when a mission site location is clicked. Students are currently selecting video clips to be digitized for publication on this page.

The Aquanaut Program Monitoring Pages. The AP Monitoring Program research theme serves as the data source for the AP Monitoring Program Pages. This group of Phase II students is developing a WWW page that is divided into two main components: the Heavy Metal Monitoring Page and a page dedicated to comparative data sets.

Heavy Metal Monitoring Page. The Heavy Metal Monitoring Page will present the information on the first core AP research topic. The page will be structured similar to a scientific paper, including: (1) an introduction that contains background information on the topic of monitoring, (2) a materials and methods section that describes and shows the technologies, animals, and techniques used in the Monitoring Program, (3) a results section that lists past and present results in spreadsheets and graphs, and (4) a discussion that describes what the results mean and what more needs to be done.

The introduction contains links to federal and state agencies with ongoing monitoring programs, including NOAA's National Status and Trends program, the Environmental Protection Agency, and the Massachusetts Water Resources Authority. The methods section links to sites describing the submersibles and ROVs used to collect the indicator organism, the sea scallop. Links to sites describing the biology of the sea scallop and images of scallops on the seafloor will soon be included. The results section contains data tables and summary graphs of the heavy metal data collected by AP between 1991 and 1995. The discussion section is under construction.

Comparative Data Sets. The concept of comparative data sets is relatively new. The amount and variety of hands-on data being collected by

Aquanaut teachers and students provide an excellent opportunity to share data with other marine and aquatic education programs nationwide. The objective of the comparative data page is to have students collect, analyze, and present data from the marine environment that have a high degree of universality associated with them, i.e., variables that can easily be compared from one location to the other. The data will be published in raw form as part of the Monitoring Program page with suggested ways to analyze and visualize the results. Comparative data are an excellent pedagogical tool that stimulates inquiry-based learning to address the fundamental causes of observed differences or similarities in the data obtained. Another important aspect of locally collected data is that they are relevant and important and, therefore, of interest to the students. At the same time, local data provide opportunities to discuss the causal factors creating the differences and the relative importance of the variables and values measured to other water bodies. Teachers work with Phase II students to integrate these locally collected data with existing data sets that can be compared to others around the world. Among the data parameters measured by the students are turbidity, CTD profiles, plankton (species composition, diversity), and nutrients (nitrates, ammonia). The information collected for 1996 is currently being collated and organized for publication on the AP AquaNet Monitoring home page.

IMPACTS OF THE AP

Impacts on Student Participants

The following illustrates the effectiveness of the AP-developed inquiry-based methodology on the student participants. An annual assessment and evaluation of the program is conducted in late summer each year that serves to gauge the effectiveness of the program and to maintain contact with student alumni. To date, a total of 450 alumni (of a total of 855 participants) have been sampled over the past 5 years. Overall statistics indicate that 89% of the alumni are now attending a college or university. The remaining 11% represent students who are now in the work force, technical schools, or are still attending high school. Twelve percent of those who are pursuing an advanced education stated that the AP influenced their decision to attend college; all others said that they had already planned to pursue a higher education prior to participating in the program. Most significantly, 70% of those attending college said that the AP directly and positively affected their choice to pursue a career in the sciences, engineering, or mathematics. This represents a marked impact of this program and a high degree of achievement in meeting the objectives of the AP.

With regard to environmental awareness, an overwhelming figure of 98.8% of the alumni stated categorically that the AP directly influenced their interest and understanding of environmental issues, especially with regard to the marine environment. This is a remarkably consistent achievement of the environmental education objectives of the program.

All of those surveyed stated that the most memorable experience of the AP was the hands-on advanced technology component, such as piloting the ROVs, submersible dives, and on-board ship activities. All stated that they would do it again without hesitation, though reports of the personal gains derived from this program were varied throughout all of the students surveyed. In every case the respondents indicated the program was an intensive, challenging yet enjoyable learning process.

The accomplishments of student Aquanauts in state science fairs is also noteworthy. 1996 was the third year that Phase II Aquanauts have presented the results of their projects in state science fairs with excellent results. In 1994 three Phase II Aquanauts placed second as a team in the grades 9–12 biology category, and in 1995 six students competing in the grades 9–12 physics category placed second in team competition and one finished in a two-way tie for third place in the single entrant category. In 1996 three teams placed, including second- and third-place finishes. For three years running special awards have been garnered from the Acoustical Society of America, United Technologies, U.S. Department of Agriculture, and Mystic MarineLife Aquarium.

Impacts on Teacher Participants

Responses on questionnaires from teachers have been a primary input for the improvement and evolution of the AP. Teachers have said they were reintroduced to the rigors of field research, which was markedly different from the experiments conducted in school laboratories where little, if anything, ever fails. They were able to see, and later discuss with their students, the complexities of performing research in the field, as well as the dependency of such large efforts on personnel and sophisticated equipment. All indicated they benefited from the chance to experience the application of state-of-the-art technology first-hand. All teachers agreed that not only did the AP allow them, as educators, to refocus and enhance their own knowledge, but also that it gave them new ideas and the means to achieve more dynamic science programs within their schools. It allowed them and their students to interact with the scientists as people and not stereotypes derived from textbook or popular accounts of science and scientific discovery.

The AP has also been successful in providing teachers with materials to produce state-of-the-art teaching materials, to make use of the WWW and to bring those resources and materials to other teachers and students over long distances. The AP home page (http://www.ucc.uconn.edu/~wwwnurc/apintr.html) provides an introduction to the structure of the program, its research themes, and information on enrolling in the program.

Linkages with Other Agencies

The AP has formed productive linkages with many institutions and agencies during the past 4 years. Among the collaborators who have provided technical expertise, in-kind cost sharing, and logistical support are federal agencies, state agencies, and private institutions. Among the federal agencies are: NOAA's National Marine Fisheries Service Northeast Fisheries Center, Woods Hole, MA; NOAA's National Marine Fisheries Laboratory, Milford, CT; NOAA's National Marine Sanctuaries Program, Plymouth, MA; the Environmental Protection Agency (EPA), Narragansett Laboratory; EPA's *Ocean Survey Vessel Peter Anderson*; the U.S. Geological Survey, Quissett Laboratory, Woods Hole, MA; the U.S. Coast Guard Research and Development Center, Groton, CT; the Naval Undersea Warfare Center Detachment, Newport, RI, and the U.S. Army Corps of Engineers, Boston, MA. The state institutions assisting the AP include: Maine Department of Marine Resources, West Boothbay Harbor, ME; The University of Connecticut's Marine Science and Technology Center and Marine Science Department, Groton, CT, and the University of Maryland, Columbia, MD. The private institutions lending support to the AP include: Project Oceanology, Groton, CT; the Mystic MarineLife Aquarium, Groton, CT, the Norwalk Maritime Center, Norwalk, CT; the New England Aquarium, Boston, MA, and Rescon Associates, Columbia, MD. The AP is run in collaboration with Gordon College, Wenham, MA, which provides room, board, laboratories, and classrooms for all work ashore during July and August each year.

Public Outreach Impacts

The AP has also provided an excellent venue for promotion of the NURP and *in situ* science in the public media. Aquanauts have presented their work to local television and radio stations as well as to other agencies such as the Connecticut Department of Environmental Protection. In 1993 the AP received coverage from CNN's Science and Technology for work conducted on Stellwagen Bank. In 1994 a non-NOAA funded, international liaison between U.S. Phase II Aquanauts and Canadian students on the Great Lakes was covered by numerous Canadian and U.S. television stations. In 1995 local Connecticut NBC news covered the work of the hearing-impaired Aquanauts in the SBNMS with a three-part series on nightly news. For 1996 a *National Geographic* writer developing a story on National Marine Sanctuaries was on board the AP cruise to SBNMS and will include the AP to be noted as a part of NOAA's education activities in his article on research and education in National Marine Sanctuaries.

SUMMARY

Since 1988 the AP has been a dynamic educational enrichment program that has sought to link the research enterprise more closely with the education process. In so doing, the program has relied on three key elements. First is the ability of advanced technologies to bring the discovery of the oceans to teachers and students. The program has linked researchers with educators to form dynamic partnerships that have resulted in mutual benefit. The program has immersed teachers and students in the study of real-world issues, in which their discovery may be the first of its kind or a contribution to our understanding of complex issues affecting our coastal resources. The program is attempting to translate the excitement of discovery and research to others via the multimedia capabilities of the WWW.

REFERENCES

I.G. Babb and P. Scheifele, "Marine Science Education Using Low-Cost ROV Technology," *Marine Technology Society Journal* 26(4)(1992):73–78.

H. R. Hungerford and T. L. Volk, "Changing Learning Behavior Through Environmental Education," in: *Environmental Education: A Component of Sustainable Development* (UNESCO Report, United Nations, 1990), pp. 8–17.International Association for the Evaluation of Educational Achievement. 1989. Student Achievement in Science in Thirteen Developed Nations. Washington, DC.

CHAPTER 6

EARTHWATCH

Science Education Worldwide

Rachel Nixon

EARTHWATCH provides individuals from across the globe with an opportunity to experience science firsthand while tackling some of the Earth's most pressing problems. Since 1972, over 50,000 EARTHWATCH volunteers have worked side by side with leading scientists and scholars to sustain the world's environment, monitor global change, conserve endangered habitats and species, explore the vast heritage of our peoples, and foster world health and international cooperation. EARTHWATCH has mobilized over 1920 projects in 118 countries and 36 states in the last 25 years. The mission of EARTHWATCH is to improve human understanding of the planet, the diversity of its inhabitants, and the processes that affect the quality of life on Earth.

In the field, EARTHWATCH scientists and volunteers become students of the Earth's processes. They immerse themselves in scientific inquiry with the goal of piecing together observations and data to arrive at substantiated discovery. Step by step, volunteers and scientists alike ask questions, small and large, about the state of the environment and its peoples. Day by day, EARTHWATCH research teams uncover history, discover new species, and establish natural reserves (Fig. 1).

In its effort to promote understanding while working in harmony with the planet, EARTHWATCH offers a unique model for science education. Through hands-on participation, EARTHWATCH volunteers, members of the public from around the world, become empowered scientists. Spending an average of 2 weeks in the field as an integral part of a research team, these individuals acquire skills important to scientific observation, data collection, inquiry, and teamwork.

Rachel Nixon, EARTHWATCH, 680 Mt. Auburn Street, Box 9104, Watertown, MA 02272; rnixon@earthwatch.org; http://www.earthwatch.org.

Figure 1. EARTHWATCH volunteers and scientists work together to unravel the mysteries of *Uncompahgre fritillary*, North America's rarest butterfly. (Photograph: Carolyn Hine)

Volunteers discover the tools of technology and become experts on the topics at hand. It is this investment in research, the connection and the passion, that drives the learning and enlightens the world.

As teachers head to the field on EARTHWATCH projects each year, students in schools worldwide reap the benefits. The direct exposure, the participation in field science, and the involvement in discovery generate a spark of excitement that inspires a renewed model for teaching and learning back in the classroom. Teachers of all levels and disciplines become students in the field, relearning the components of inquiry, understanding, and mastery of a scientific topic. They return to their students refreshed and invigorated with new energy and expertise. Abandoning outdated lessons, these teachers strive to instill a quest for discovery and a love of learning in their students.

High school English teacher Merey Grearson headed to Bahia in the summer of 1996 to study black sea turtles with EARTHWATCH scientists and volunteers. Her journey of discovery revitalized her passion for learning and reconfirmed her commitment to teaching. She writes:

> Perhaps the greatest gift this expedition gave me was the reminder of why I teach....Being in Bahia de Los Angeles, a world so utterly unfamiliar to me, and helping to do research in an area that was as unfamiliar as the country around me, I suddenly remembered how much I love to learn, how much I ache for understanding and revel

in new experiences. I felt the familiar thrill of everything making sense, of putting new skills to use, of figuring out a problem, of asking questions. To be in a situation to reach students, to see the light bulb go on above their heads, to hear a student say, 'Oh, I get it!' That is why I teach.

Gregory Bisbee, a biology teacher from Wisconsin, has participated in several EARTHWATCH expeditions. In July 1996 he worked with research scientists Christopher Pague and Aaron Ellingson on the rugged alpine slopes of Colorado to help save an endangered species of butterfly. He writes:

> I caught the excitement and enthusiasm of doing research in the field. When the field staff talked about the research, you could see and hear the excitement of discovery that I attempt to convey to my students in the biology classroom. It is this sense of excitement that draws people to the sciences....EARTHWATCH is one of the best ways that I have found for renewing my excitement and love for science.

Instilling this sense of excitement in teachers is the key to successful science education. EARTHWATCH research scientists serve as models for educators; they inspire teachers to be positive mentors back in the classroom.

Educators return from the field with specific skills that they wish to teach their students. Second grade teacher Katherine O'Brien reflects on her experience excavating an Etruscan farm in Italy:

> Taken as a whole, this expedition experience opened my eyes. The work ethic that I witnessed in the Italian archeological team was unsurpassed. This is the main concept that I wish to share with my students. Namely, that nothing of any value in life is achieved without hard work, goal-setting, and follow-through to completion of whatever the task may be. This invaluable EARTHWATCH experience changed my approach to teaching science by not only heightening the need for students' critical thinking and observational skill but also the importance of team-work and cooperative effort to reach a scientific conclusion.

As illustrated by these three teachers, EARTHWATCH is a leader in the field of experiential education. With the help of private donors, foundations, and corporations, EARTHWATCH provides fellowships for more than 200 teachers each year, enabling them to join expeditions and experience science directly. To date, more than 2500 teachers have received career training and inspiration on EARTHWATCH expeditions through the fellowship program. In March 1996, the EARTHWATCH Education Awards Program received a 5-year endorsement from the National Science Teachers Association (NSTA). This NSTA vote of support is a powerful signal to schools and the program's donors that EARTHWATCH delivers important hands-on field training and classroom resources to teachers.

EARTHWATCH EXPOSES TEACHERS TO TECHNOLOGY

EARTHWATCH expeditions and educational programs offer exciting opportunities for teachers to work with technological products. In the field, scientists train volunteers to use low- and high-tech equipment and, more importantly, to understand the function of each tool. Then, in the quest for raw scientific discovery, educators work with the equipment directly, gathering data and drawing conclusions.

On the waters of Puget Sound in Washington, for example, principal investigator Kenneth Balcomb explains the use of the hydrophone and Global Positioning System (GPS) in the ongoing study of the *Orcinus orca* population in this region. Volunteers listen in awe as the hydrophone relays the underwater conversation of a pod of killer whales and observe as the GPS displays the latitudinal and longitudinal position of the whales. Technology advances science to seemingly infinite realms.

On the *California Wildlife* project in the Golden Gate National Recreation Area, Geographic Information Systems (GIS) expert Marcia Seminoff-Irving works with Dr. Judd Howell and EARTHWATCH volunteers to record the location of the plants and animals in the park, so as to develop a plan for harmonious coexistence of human visitors and wildlife. A similar GIS system is used to track the location of cheetah on an EARTHWATCH project in Namibia. Under the direction of Laurie Marker-Kraus, volunteers radio-track cheetahs by airplane and enter data pertaining to the daily patterns of these animals into a GIS computer program (Fig. 2).

Figure 2. The radio collar on this cheetah allows EARTHWATCH scientists and volunteers to gain insight into the movements and distributions of this speedy mammal. (Photograph: Laurie Marker-Kraus)

EARTHWATCH also makes scientists and project information accessible via the Internet, a communications technology that is accelerating the spread of information worldwide. At the heart of EARTHWATCH's expanding approach to education is the on-line EARTHWATCH Global Classroom (http://www.earthwatch.org/ed/home.html). This interactive center, the most frequently visited of EARTHWATCH's 1000 Web pages, allows for electronic correspondence between EARTHWATCH-sponsored scientists and students in classrooms around the globe. Through e-mail, students of all ages pose questions to our world's leading scientists and eagerly await responses. This technology breaks down the barriers between classrooms and the seemingly distant scientific world.

The EARTHWATCH Global Classroom also provides information from expeditions for use in schools. Project features highlight all aspects of expeditions: maps, photos, field notes, and journals from volunteers, interviews with the scientist, project data, sounds from the field, and teacher-developed lesson plans. EARTHWATCH's curriculum specialist Meg Warren works with teachers, primarily through e-mail and telephone, to post the latest information on the EARTHWATCH Global Classroom.

Lastly, new technology grants received by the EARTHWATCH Education Department in 1995 and 1996 empower teachers to take the controls using video cameras in the field, communicating through e-mail, posting information on the Internet, and developing collaborative projects using groupware such as Lotus Notes.

EARTHWATCH PROJECTS: CASE STUDIES

Testing Technology: The Total Station and Measuring Tapes

EARTHWATCH project: *Modeling Forest Canopies*
Principal investigator: Dr. Jiquan Chen, Michigan Technological University
Research site: Changbaishan Natural Reserve, Jilin Province, China

Scientists have recently begun to explore forest canopies throughout the world. Most studies to date have focused on the diversity of the canopy itself, but a new line of research strives to understand how canopy composition and architecture affect the entire forest ecosystem. With the help of EARTHWATCH volunteers, Dr. Jiquan Chen is working to develop a three-dimensional computer model of a 300 × 300-m plot of virgin forest in the spectacular Changbaishan Natural Reserve in China. Achieving the goals of this labor-intensive project would give reserve managers and foresters ways to predict the effects of two major shapers of forest architecture (fire and wind throw) and to test alternative practices (thinning or harvesting), without having to fell trees.

Dubbed "the Forest Sea" in Chinese because of its rich and unique wildlife, the Changbaishan Natural Reserve lies in the mountains bordering North Korea, roughly along the same latitude as Connecticut. The huge reserve encompasses 734 square miles, and within its borders, Chinese scientists have documented 50 mammals, 200 birds, 1250 vascular plant species, and 87 ferns. Many-layered with canopy gaps, ancient downed logs, and branches thickly coated with epiphytes, this is a classic old-growth forest. Pine, larch, spruce, and fir predominate, and over 300 inches of rain per year keeps the forest incredibly lush. With luck, an EARTHWATCH volunteer on the *Modeling Forest Canopies* project might glimpse a black bear, marten, sika deer, wild pig, or cougar.

Heading to the field in the summer of 1996 for a 2-week research experience, EARTHWATCH volunteers worked with Dr. Chen and his international support staff of botanists and foresters to lay out the sample plot. Volunteers were responsible for tagging trees with stems greater than 6 cm, measuring tree and canopy dimensions, and using a computerized "total survey station" to record the location of each tagged tree. The volunteers also sampled understory plants in 1-m^2 plots along the study area's two transects. Several research devices were used in the gathering of data: a diameter tape, clinometer, compass, and the survey station (a GPS).

Elizabeth Revenko, a middle school science teacher from California, participated in Dr. Chen's project in July 1996. While on site in China, she worked directly with high-tech equipment. Elizabeth writes:

> I spent most of my time working with the Total Station, an expensive piece of equipment so named for its ability to both accurately measure the spatial location of trees and store the data in the field via a data logger for later downloading into a computer back at the research station. The Total Station uses a laser beam to measure distance to an object which the data logger translates into X, Y, and Z coordinates....This aspect of the project was particularly interesting to me, both personally and professionally. I was lucky enough to be afforded an opportunity to work closely with the most technological aspect of the project, which gave me a good overview of the project design and a good understanding of how technology was intended to be used to make the tree plotting easier and more accurate.

From day one of the project, Elizabeth was given the guidance needed to comprehend the purpose of data collection using the Total Station. She worked closely with Chinese researchers to understand the high-tech device (Fig. 3). The scientists served as mentors in this process of learning. She writes:

> It was neat to work with the Chinese researchers. They were characteristically quiet, focused, and hardworking. One of the researchers endeared himself to me because he was so easy to work with. While I worked on the Total Station, I had to focus the laser on a prism which was held against the tree. He understood exactly how to hold it so that I could sight in a heartbeat. He could see when there were branches in the way or even a wispy strand of moss, and would find the only unobstructed square centimeter for me to aim the laser. And tree after tree, he worked quietly and steadily.

Figure 3. EARTHWATCH teacher fellow Liz Revenko works closely with a Chinese scientist to measure the spatial location of trees using the Total Survey Station. (Photograph: Elizabeth Revenko)

Elizabeth's work with the Chinese researcher exemplifies how EARTHWATCH projects bridge cultures, uniting individuals committed to scientific inquiry.

During this expedition, Elizabeth and the research team experienced frustration when the Total Station failed time and time again. Days of tiring data collection were lost, and attempts made by the scientists and the volunteers to repair the system were futile. Finally abandoning their unsuccessful repair efforts, the team members fell back on their most fundamental understanding of the scientific process and continued their investigations using measuring tapes. Elizabeth notes:

> This was a great experience for several reasons. It showed me the potential of technology yet underscored its limitations. Moreover, it reaffirmed my belief in the essential usefulness of knowing how to do tasks the low-tech way. I will continue to

argue that students need to know <u>how</u> to do something and that while computers are available to improve upon those techniques, it is the underlying thinking and skill that are important.

Elizabeth's experience illuminates the importance of educating oneself as to the basic function of technological devices. In the event of equipment failure, one can rely on problem-solving skills and a more fundamental, yet highly effective, process of data collection so as to continue the investigation at hand.

The work on *Modeling Forest Canopies* triggered scientific problem-solving and exposed the researchers and volunteers to high- and low-tech approaches to data collection. For Elizabeth and the other teachers on the expedition, the project underscored the values of perseverance and dedication and highlighted the importance of reevaluating methodologies in the face of difficulty. All of these values are of the utmost importance in the classroom. Elizabeth sums up her experience and its future impact on her students:

> This trip is a good reminder to me of what research is all about; why people do it, how they do it. Not just mentally organizing research, but physically doing it. It will be helpful for me to think back on this when I advise my eighth graders in their science fair projects, or really any of my students as they work through something new; to offer support when the experiment becomes frustrating and to help them try something different when the original plan does not work as expected. These are some of the most useful skills one can have.

Teachers and Students Radio-Track an Endangered Species

EARTHWATCH project: *Black Sea Turtles of Baja*
Principal investigators: Jeffrey Seminoff and Wallace J. Nichols, University of Arizona
Research site: Bahia de Los Angeles, Baja California, Mexico

Black sea turtles are among the rarest sea turtles in the world. Traditionally, this species has used the Gulf of California as its prime feeding and nursery area. In past centuries, the local Seri Indians hunted this once-abundant turtle, an easy target during its dormant winter season. At the turn of this century, white settlers began intensive harvesting of the black sea turtle for its meat, leather products, and oil. By the 1980s, the population had crashed.

Today, protected on most of their nesting beaches, few adult black sea turtles still inhabit the gulf waters. Those that do are threatened by illegal poaching and incidental capture in fish nets. Resource managers know where some of the turtles nest, but detailed information about the daily patterns of these creatures is scarce. Where and how far do adult turtles travel both daily and seasonally? Where do they congregate? What do they eat, and how does their diet change seasonally? What are the habits and needs of juveniles that are more common in the gulf?

Principal investigators Jeffrey Seminoff and Wallace J. Nichols work with EARTHWATCH volunteers to discover the basic aspects of black sea turtle life. Mornings and evenings, crews set out in boats to set up, monitor, and take down nets used to trap turtles (Fig. 4). Whenever possible, crews catch the turtles by hand. Volunteers record the exact location of the turtles using a GPS, and then the turtles are tagged. In addition, sonic and VHF radio transmitters are attached to select turtles before they are released.

EARTHWATCH volunteers are also on hand during interviews with local fishermen about former levels of exploitation, migration patterns, locations of nesting beaches, and current conservation attitudes. Seminoff and Nichols have worked hard to gain the respect of local people and have even managed to

Figure 4. EARTHWATCH scientist Wallace J. Nichols pulls up the net while EARTHWATCH volunteers check for turtles and remove algae and other debris. (Photograph: Richard Timm)

convince many members of the local community of the value of keeping the turtles alive as a prime tourist resource.

In June and July 1996, a handful of teachers from throughout the United States joined Seminoff and Nichols in the quest to understand black sea turtle life. Samantha Sonnak, a middle school science teacher, notes, "A neat fact about our team was that there were a few other teachers on our expedition and we were able to share our classroom experiences and thoughts about teaching and education in general." This sense of collaborative learning serves as a model for the development of cooperative learning groups among students.

The *Black Sea Turtle* project also emphasizes how teachers of all disciplines benefit from experiential field research. Merey Grearson, an English teacher from Massachusetts, set out on a mission to better understand science in connection with her passion, writing. She notes:

> I wanted to learn to look at things like a scientist in order to see what scientists see when they look at an object. I wanted the inside scoop on facts and statistics, details and intricacies so that what I learned might connect with something I could write about, something that would ultimately have grown from and be substantiated by some tiny germ of fact.

These eloquent words underscore Merey's desire to be exposed to the latest scientific technologies and processes. She wants to be a student of science and translate her newfound excitement back to her students. "I can and will show slides, will talk of deserts and turtles, but what can I really give them except the enthusiasm to open their eyes and look at the world around them, to go and experience for themselves?" she adds. This marks the crucial step for classroom teachers—to transfer the spark for real-life learning back to their students. Merey's excitement and energy and her love of learning will surely inspire her students to undertake similar scientific expeditions.

For Merey and the other teachers, *Black Sea Turtles of Baja* provided exposure to some of the world's leading scientists and the latest technology. According to Anne Steel, a middle school math teacher from Mississippi, Seminoff and Nichols were "the finest teachers we could have had. They taught us carefully, then trusted us fully to carry out the tasks we had been taught." Night watches included monitoring sea turtles traveling throughout the bay. Anne Steel recalls "listening with headphones at 3 A.M. to the high-pitched beeps of a turtle surfacing to breathe far out in the bay." In the dark of night, the latest technology afforded data collection and an avenue into an otherwise unknown world.

During the day, volunteers prepared select turtles for long-term sonic and VHF tracking. Anne reflects on the long-term impact that her participation in this project will have on her class back in Mississippi:

> In the afternoon we scrubbed the algae off our resident turtles' carapaces with extra vigor, some of us bidding a tender farewell to our favorite, 'Sunburst' [Fig. 5]. We traced with our fingers the precise pattern of marginal and neural scutes as we measured their dimensions with large and small calipers....Mounted at the apex of his carapace

Figure 5. EARTHWATCH teacher fellow Anne Steel bids farewell to her favorite, "Sunburst." (Photograph: Lynn Jimenez)

was a satellite transmitter we had just attached with epoxy. That transmitter would be our link over the winter to this new-found world of sea, desert and turtle, for we would be receiving satellite data reports from J and Jeff via e-mail.

Anne speaks as an expert as she describes preparing the turtle for release, and she will share this confidence and understanding with her class as they track "Sunburst" together throughout the winter. In addition to receiving up-to-date data on the turtle's latest movements, her students will pose questions to the scientists. Anne Steel sees this on-line communication as a way to introduce problem-solving and data interpretation into her math curriculum throughout the year.

Samantha Sonnak's science classes will also receive raw data from Seminoff and Nichols. She writes:

The information on the turtles' movement will be available via the Internet for my students to use and track the turtles from 3,000 miles away. We will be able to contribute

to a legitimate study and help out in our small way, by working with the scientists and
using the new technology available to keep track of the results on our own.

The Internet will break down the barriers between the classroom and field
science, and Samantha will engage her students directly in the study in which
she participated. Through this involvement, she will empower her students to
make meaningful contributions to science.

While on the expedition, Anne and Samantha decided that they would work
with their classes to raise money for a piece of equipment needed at the research
site in Bahia de los Angeles. Once the students have communicated with
Seminoff and Nichols and learned about the functions of the technological
equipment, they will choose one device that they will purchase through a
fund-raiser (Fig. 6).

This EARTHWATCH expedition illustrates how teachers from all disciplines
become empowered learners of science technology, transfer the spark of learning
to their students, and maintain a link to the scientific world when back in the
classroom.

EARTHWATCH Projects Put Scientists within Reach

EARTHWATCH project: *Community Forestry in Ecuador*
Principal investigator: Dr. Constance Becker, Indiana University; People Allied
for Nature, Ltd.
Research site: Suspiro, Loma Alta, Ecuador

During the 1995–96 school year Cheri Ewald, a research chemist in
Charlotte, North Carolina, and an EARTHWATCH team member herself, volunteered to work with fifth graders in Matthews Elementary's Science Adventure
Program on a project to explore the rain forest. With the goal of gaining
immediate access to live field research, the class decided to adopt *Community
Forestry in Ecuador* as part of EARTHWATCH's pilot "Adopt-a-Project" Program.
Through this program, Cheri and the students received the project briefing,
mission alert, a video taken by an EARTHWATCH teacher fellow, and e-mail access
to the research scientist herself, Dr. Constance (Dusty) Becker.

Cheri worked closely with the students to lay a foundation of knowledge
about the Ecuadoran rain forest, the local peoples, and the mission of Dr.
Becker's EARTHWATCH research teams. She created geography lessons using
maps from the project briefing, cultural activities about village life through the
video, and inspired the students to join the day-to-day activities on *Community
Forestry in Ecuador* through the project mission alert. The fifth graders read the
following excerpts:

> Nearby the barking roar of a howler monkey momentarily erases the twittering of
> myriad birds and makes a muffled echo through the fog that clings to these Ecuadorian
> mountain tops. You're tired, muddy, and completely exhilarated. Far below you, out
> of sight, stands the village of Loma Alta, with its fields of sweet melons, cantaloupes,

Figure 6. The radio transmitter, mounted here on the turtle's carapace, is one piece of equipment that the students are considering for their fund-raiser. (Photograph: Richard Timm)

and vegetables. Residents till the fields and tend their chickens, pigs, and goats. Tomorrow you'll be down there interviewing those villagers, just as you have been cataloguing and mapping the plants and animals of the cloud forest today.

This EARTHWATCH project is as much about people as it is about the forest. Here in the coastal mountain range of Ecuador, people hold communal rights to the land. The villagers here have invited community forestry ecologist Dr. Dusty Becker and EARTH-WATCH volunteers like yourself to help them assess the biological richness of their cloud forest and to help them devise a way of preserving the forest and integrating their activities on the land. It's exactly what sustainable development and enlightened resource management are all about, and you're on the cutting edge of this vital issue. [Fig. 7]

Ecuador is dramatically divided by the Andes. On the western slopes, down to the sea, the land is dry, while to the east stretch the jungles of the upper Amazon basin. Loma Alta is in the southwestern coastal foothills. Major watersheds created by these mountains are owned by indigenous communities. About two-thirds of comuna Loma Alta lies below 300 meters; here corn, peppers, beans, tomatoes, and melons are cultivated. Above 300 meters and on steeper slopes there are plantations of a palm used for making Panama hats, remnant coffee plantations, and stands of moist tropical forest. These forests are literally holding the soil in place, keeping the streams and rivers from taking the mountains out to the sea. While parts of the forest have been disturbed by peasant farmers, intact portions are thick with epiphytes, gigantic leaves, and buttressed trees. The booming calls of howler monkeys are heard at sunrise and sunset, while toucans and other birds are active through the day.

Figure 7. EARTHWATCH teacher fellow Ken Kopperl measures water combed from mists wafting in from the nearby Pacific Ocean. (Photograph: Ken Kopperl)

Through this intensive multimedia introduction to Dusty Becker's project, the fifth graders at Matthews Elementary learned about geography, field science, and cultural attitudes and actions. They were exposed to new species, new noises, new feelings, and global issues such as sustainable development and resource management.

Having gained this new expertise, the students worked diligently to compose questions for Dr. Becker. The climax of their learning took place in the electronic mail correspondence with Dusty. Student inquiries ranged from "How long have you been studying rain forests and visiting them?" to "What things do the people of Suspiro eat?" Below are excerpts from Dusty Becker's responses to the fifth graders' questions, which are posted on the <u>Ask the Scientist</u> section of the EARTHWATCH Global Classroom.

Kristen, Ms. Allen's Class. How long have you been studying rain forests and visiting them?

Dusty: I first visited a rain forest in 1962 when I was 7 years old. My family was living in Chengmai, Thailand because my dad worked for the World Health Organization. My Brownie Scout Troop went on an overnight campout at a missionary retreat in the rain forest. We saw tiger and some amazing birds and butterflies! What I remember to this day, was how people used bamboo from the forest as pipes to transport water down the mountain. Bamboo is very useful.

Figure 8. EARTHWATCH scientist Dusty Becker at work in the field. (Photograph: Peter Tyson)

Ashlyn, Mrs. Holman's Class: How did you get to be the "leader" of the mission in the cloud forest and village?

Dusty: The project was my responsibility because it was largely my idea. As a principal investigator, one takes on the risk of success and failure. I had applied for funding to do this project, so I was the leader or coordinator (better word for it, I think). Many others helped me make the project work and sometimes they played leadership roles. I was not always the "leader," especially when it came to speaking Spanish. I relied on many others for making the project a success. [Fig. 8]

Mary, Ms. Dimitroff's Class: What kind of food did you eat in Ecuador?

Dusty: We had great food. Ecuador has lots of great fresh fruits and veges, so plenty of healthy meals. I ate fish, chicken, shrimp, bananas, tomatoes, soups, beans, rice,

cassava....You should ask to see the answer I sent to Matthew in Ms. Allen's Class,
where I described the food that people eat in Suspiro. In the cloud forest camp, we ate
too much tuna fish and pasta. I was pretty tired of tuna by the end of the trip.

The thoughtfulness of the questions reveals the investment and enthusiasm of the students, and Dusty's descriptive responses illuminate her dedication to science education and her interest in the young students. She recalls her experiences as a child, a key bridge to the fifth graders' world, and stresses the importance of teamwork in scientific endeavors, an important message for the students to grasp and apply to their own classrooms.

Inspired by the direct contact with the scientist and the field methodologies explained on the video, Cheri and the students conducted parallel research projects in the Matthews Elementary schoolyard. Following the model of the EARTHWATCH volunteers working to inventory the cloud forest in Ecuador, the fifth graders undertook biodiversity studies of their local surroundings. They catalogued plants and animals around their school, measured trees, and then compared their findings to the data collected in the Ecuadorian rain forest.

All six fifth-grade classrooms at Matthews Elementary participated in the Ecuadoran rain forest project, reading the background information, studying maps, watching the field video, posing questions to Dusty Becker, and conducting scientific research in their own backyard. Cheri Ewald, the coordinator of this project, notes: "The kids got a real taste of what field research can be like. They learned about the logistics of pulling it off and they got the sense that the scientist handling the research is a regular person."

In addition, these students, and the entire educational world, have access to other primary documents from teachers who volunteered on Dusty Becker's project. Educators and students can click on Photos and view a photo essay comprised by Joan Horn, a middle school language arts teacher from New Jersey. They can "visit" the dusty village of Suspiro and view students, like themselves, helping Joan to prepare educational posters to be used in an upcoming village assembly. From there students can click on Field Notes to read a letter written on July 20, 1995 from Mirah Skillman to her students at Grace Rogers Elementary School in New Jersey. The possibilities for learning are endless as one peruses these primary sources on the EARTHWATCH Global Classroom.

Distance Learning: Connecting Teachers in the Field and the Classroom

EARTHWATCH project: *Amazon Katydids*
Principal investigator: Dr. David Nickle, National Museum of Natural History, Smithsonian Institution
Research site: Amazon Basin, east of Iquitos, Peru

Katydids, known for their cricketlike nocturnal singing, can be found in forests throughout the world. The 360 species of katydids identified so far in

Figure 9. "Spiny Devil" katydid. (Photograph: David Nickle)

the Peruvian Amazon come in a startling array of colors, shapes, and sizes. As bats, birds, monkeys, rodents, snakes, spiders, wasps, and even ants dine on katydids, these crepuscular choristers have evolved amazing adaptations to elude predators (Fig. 9). Camouflage artists, katydids mimic twigs, pieces of bark, leaves, lichen, even bird droppings as they rest, otherwise defenseless, during the day.

Since 1987, Dr. David Nickle has been studying these intriguing insects with the help of EARTHWATCH volunteers. In the summer of 1996, two EARTH-WATCH-Bell Atlantic teacher fellows headed to the Amazon, along with EARTH-WATCH's Director of Education Tally Forbes and seven other volunteers, to collect katydids, videotape field methodologies and activities, share teaching ideas, and experience the Amazon.

The Bell teacher fellows on *Amazon Katydids*, Bob Frostick and Penny Jenkins, are 2 of 11 teachers who received EARTHWATCH teacher fellowships through the Bell Atlantic Foundation in 1996. In addition to traveling to the Amazon to study katydids, Bell-sponsored teachers joined EARTHWATCH teams to radio-track deer in the Shenandoah Valley in Virginia, excavate Brimstone Hill on St. Kitts, study the forest islands of Bolivia, explore the effects of acid rain on trees in Bohemia, and investigate the carnivore species in a Mexican

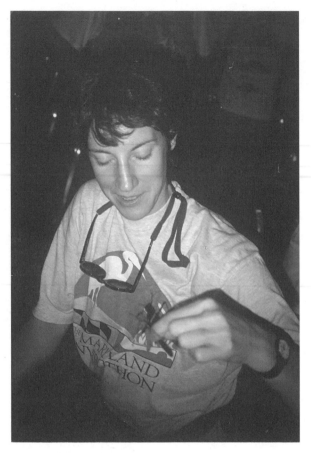

Figure 10. EARTHWATCH teacher fellow Penny Jenkins observing her katydid catch. (Photograph:
Tally Forbes)

forest. In its second year, the theme for the Bell Atlantic grant is "Building
Collaborative Projects for Networked Schools." During the 1996–97 school year,
the 11 teachers, as well as 2 EARTHWATCH-Lotus fellows, are participating in a
pilot project using Lotus Notes as the medium for sharing data, lesson plans,
pictures and collaborating on classroom projects based on their EARTHWATCH
experiences.

The emphasis on networking began immediately on the Bell Atlantic grant
renewal in the spring of 1996. The EARTHWATCH-Bell application was sent via
electronic mail to the Bell Atlantic recruiting contacts in Delaware, Maryland,
Washington, D.C., Virginia, West Virginia, Pennsylvania, and New Jersey. Updates were sent back and forth weekly. Within only 2 weeks, the EARTHWATCH

Education Department had already received several teacher applications via e-mail. When selecting candidates, EARTHWATCH and Bell Atlantic looked for teachers with an adventuresome spirit, a commitment to experiential learning, and an interest in expanding the use of technology in their classrooms. Penny Jenkins sent her application via electronic mail and explained her interest in the project and the subsequent carry-over into her classroom:

> The 'hook' of the Amazon experience, which I can readily share with photographs, tape recordings, video and computer graphics as well as the on-line research, will enhance, motivate and excite students into working to become scientists themselves. Students will gain experience conducting research and using the Internet to connect the classroom to the real world.

Once the teachers were chosen and their projects assigned, concrete preparations began immediately. Bob Frostick, Penny Jenkins, and Tally Forbes communicated weekly via e-mail about their upcoming Amazon adventure. Questions such as "What kind of boots are you bringing to protect yourself from the killer ants?" and "What shots are you getting?" bounced back and forth among this threesome.

On arrival in Peru, Bob, Penny, Tally, and the other EARTHWATCH volunteers met Dr. Nickle, the research staff, and the katydids. While becoming accustomed to the surroundings, they dodged stinging insects and plants and soon realized that they had to look up as well as down for protection. Heading out into the rain forest at night, when katydids are active, the team members searched for and collected the insects and recorded behavioral observations (Fig. 10). In the morning, volunteers processed the data from the night's work.

Day by day, Bob and Penny collected video footage and photographs, recorded their insights in journals, and considered ways in which their experience could be translated back into their classrooms in West Virginia and Maryland (Fig. 11). They knew they would never forget the walks along the canopy walkway, the splash into the Amazon, the visit to the Peruvian schoolhouse, and the face-to-face contact with a boa.

Bob, Penny, and the rest of the Bell teacher fellows returned from the field refreshed, excited, and beaming with new ideas. On September 20, 1996, the Bell Atlantic and Lotus teacher fellows gathered for a training workshop on the EARTHWATCH-Lotus Notes application, created specifically for these educators and their respective EARTHWATCH projects. The teachers learned how to navigate this interactive, collaborative platform and how to post their field journals, maps, slides, project information, and curriculum ideas. In addition, they experimented with the discussion groups, an avenue for both teacher and student communication among the various schools.

Back in their classrooms along the East Coast, the teachers are now learning the intricacies of their schools' computer systems and are working with

Figure 11. EARTHWATCH teacher fellow Bob Frostick videotapes scientific methodologies. (Photograph: Tally Forbes)

technology coordinators, and the support team from Bell Atlantic, to load the Lotus Notes software and prepare for the necessary connections to the server which houses the EARTHWATCH-Lotus Notes application. They are learning the complex world of computer lines, modems, and software, as well as the processes of data entry and replication.

Through Lotus Notes, the possibilities for collaboration are endless. The application serves as an interactive resource library in which teachers and students can post and access information about the six different EARTHWATCH projects visited by the Bell Atlantic and Lotus fellows. While working in the Notes application, teachers and students load information into the project databases, view other entries, comment on topics posted by other teachers, communicate via electronic mail, pose questions for discussion groups, and collaborate on classroom projects.

EARTHWATCH sees great potential in this pilot project. After spending 2 weeks at a research station tackling some of the world's most pressing environmental issues, these teachers are now exploring, firsthand, the rapidly expanding world of technology. EARTHWATCH, Bell Atlantic, and Lotus have clear goals for this collaborative project: to spark enthusiasm about science among teachers and students, to encourage experimentation with technology, to promote the sharing of knowledge and resources, and to consider the capabilities of the Lotus Notes software.

While waiting for the appropriate technology necessary to upload the EARTH-WATCH-Lotus Notes application at his school in West Virginia, the "virtual" Bob Frostick created his own homepage titled "An Amazon Adventure" (http://168.216.210.13/amazon/). While visiting his site, one can view pictures from the expedition, maps of the research area, and reports and artwork created by his eighth grade students at Horace Mann Junior High School, as well as students from other schools nationwide. His students have researched plants of the Amazon rain forest, such as the philodendron and rubber tree, and posted their findings along with photographs from their teacher's EARTHWATCH expedition.

In an electronic mail message to the Director of Education Tally Forbes, his teammate on *Amazon Katydids*, Bob describes another activity and its posting on his Web site: "I currently have all the kids out grabbing grasshoppers, katydids and the like and will be putting up pictures of the West Virginia katydids as well." Bob has taught his students the data collection methods that he learned in Peru. Once the insects have been gathered, his students will engage in a comparative study of the West Virginian katydids and the Peruvian species. As with all research, there are often obstacles along the way. Bob writes: "Someone moved the cover board and I had a major jail-break. Grasshoppers and katydids are running the room and the hall!" It is this enthusiasm and commitment that will carry Bob, the rest of the teachers, and their students through the next phase of loading information into the Lotus Notes database and collaborating from school to school.

Low-Tech Tools, High-Tech Learning

EARTHWATCH project: *Homol'ovi I*
Principal investigators: Dr. E. Charles Adams and Richard Lange, Arizona State Museum, University of Arizona
Research site: Winslow, Arizona

> In the sixth grade in Weymouth, our social studies curriculum emphasizes the study of past civilizations. The students research culture, customs, values, religion, and government. In science, the theme continues to include the study of archaeology, artifacts, what happens at a dig site, and how to interpret the information that is collected. The Homol'ovi Project would be a natural bridge to bring history and science to life in a city school.

In her application for an EARTHWATCH-NYNEX teacher fellowship, Karen Travers explains how her participation on the EARTHWATCH project *Homol'ovi I* will enhance her curriculum and emphasize interdisciplinary studies. Karen is one of three teachers from the South Shore Educational Collaborative in Massachusetts who received a grant from NYNEX in the summer of 1996. Her goal of connecting city students to field research represents a primary focus of the EARTHWATCH Education Program. Through the use of technology, as well as hands-on science activities, elementary, middle, and high school students in cities nationwide are benefiting from their teachers' EARTHWATCH experiences.

On June 20, 1996, Karen Travers, Kathy Womersley, and Nancy Guenther, the three EARTHWATCH-NYNEX teacher fellows, joined Jim Lehane of NYNEX, Mike Savage of the South Shore Educational Collaborative, and the EARTH-WATCH Education Support Team for a preexpedition workshop. The group met in the computer lab at Braintree High School, surrounded by 25 computers logged on to the Internet. During the meeting, the goals of the Nynex grant were discussed: to provide curriculum content through firsthand field research, to enlighten and refresh teachers, to familiarize teachers and students with the Internet and Hypertext Mark-up Language (html), and to encourage collaborative projects among schools. In thinking through this mission, Karen eloquently states:

> The Homol'ovi Project is a tremendous opportunity to connect today's students to yesterday's culture through technology....we can put our daily research on-line by maintaining a web page. When it's complete, it should contain pictures, movie files, sound files, as well as all the details of the trip. With that complete, the students in the classroom will work to familiarize themselves with accessing a specific web site, finding their own way around the Internet and understanding site links....This would be the beginning of using the Internet as a standard research tool in the classroom.

With these goals in mind, the three teachers packed their bags with sun hats, old sneakers, water bottles, and video equipment and headed to Winslow, Arizona, to begin their journey into the past.

The mission of the *Homol'ovi I* project is to enhance the understanding of early Native American history in the Southwest through archaeological research. Named for the mounds in which they resided, the Homol'ovi peoples changed dramatically between A.D. 1250 and 1400. Originally settling in small communities consisting of pit houses, these early Americans began to construct brick and mud settlements, many of which grew fivefold during this expansive period. Today, six of these large settlements remain along the Little Colorado River near Winslow, Arizona. Hopi Indians, who reside 60 miles to the north, consider these pueblos to be ancestral homes.

Karen, Nancy, and Kathy joined the archaeological team, in its tenth season with EARTHWATCH volunteers, to excavate Homol'ovi I, a 500-room pueblo that was occupied for a longer period of time than any other settlement in the area. On July 7, 1996, Nancy writes in her journal:

> After a fan-shopping trip to Walmart, the orientation begins. The wind really kicked up and it was difficult to see, let alone breathe. One can see why the rooms were filled in with sand. The Little Colorado River is almost non-existent....Potsherds and lithics are everywhere inviting you to pick them up.

The site was bustling with EARTHWATCH volunteers, the research staff, and the occasional tourist group—all yearning for a deeper understanding of the Hopi

Figure 12. EARTHWATCH teacher fellow Karen Travers screens for artifacts in East Plaza at *Homol'ovi I*. (Photograph: Karen Travers)

origins. Signs posted at Homol'ovi alerted visitors to "Take only pictures. Leave only footprints."

After the dusty tour of the site and an introduction to the identification of lithics, ceramics, and bone, the EARTHWATCH volunteers set out with their crew chiefs to begin excavation in different sections of the pueblo. Karen worked on the East Plaza (Fig. 12), Nancy on the 100-room block, and Kathy on the kiva.

Day by day, the teachers recorded their experiences on videotape, slide film, and in journals. They collected pieces of Hopi ancestry and collaborated on curriculum ideas for their classrooms. They were inspired by the unique discoveries of the site and reveled in the piecing together of the past. They became experts on archaeology and field research methodologies. Throughout the project, with her eighth grade social studies classes in mind, Kathy was particularly intrigued by the tools. She writes:

> The actual archeological experience was rather revealing. First of all, it never occurred to me that their use of tools involved everyday items that could be purchased at a hardware store. It was interesting that the tools, such as a dust pan or a paint brush, had rather different uses on an excavation site.

It was through this use of low-tech tools in the field that Kathy was inspired to design a high-tech lesson plan, on the *Homol'ovi I* Web page within the EARTHWATCH Global Classroom, on her return from the field.

In two successive workshops at EARTHWATCH Headquarters in the fall of 1996, the NYNEX teacher fellows met with the EARTHWATCH Education Support Team to share their stories and archaeology expertise and to begin work on their EARTHWATCH Global Classroom Web site. Slides, video footage, and photographs were accompanied by tales of uncovering "The Pot," learning from the crew chiefs, and attending a Hopi ceremonial dance. The wealth of knowledge and wisdom gained by these teachers while on *Homol'ovi I* impressed the EARTHWATCH staff and no doubt inspires their students.

The second workshop provided the teachers with hands-on "html" training and resulted in the posting of images, journals, and lesson ideas on the Web the very next day. EARTHWATCH's curriculum specialist and Global Classroom developer Meg Warren led the teachers, and other EARTHWATCH staff members, through the basics of "html." Instantly, the teachers were up-and-running on the creation of the *Homol'ovi I* Virtual Field Trip. Each teacher worked eagerly on his or her excavation site, using codes such as <center>, <p>, <img align=left>, and <b>, to instruct the computer to present text and images in a particular manner when uploaded on the Internet. Cheers of elated joy rang out in the EARTHWATCH Education Department when files were opened and the magical presentation of the data and pictures appeared. In addition, the teachers learned how to create links to each other's work and develop interactive lessons that could be used by their students.

Kathy, for example, translated her interest in the excavation equipment into an activity where students match tools with their corresponding functions. When the correct tool is chosen, the participant is linked to an image showing the tool in action on *Homol'ovi I* (Fig. 13). Back at her school in Scituate, Massachusetts, Kathy uploads her lesson, tries it out, and then corresponds with Meg Warren at EARTHWATCH:

> Hi Meg. Cool, cool, cool. I love it. Thanks so much for putting me into the 90s! On the activity Tool Time, the first answer came up really quickly (30 seconds), the next two, (#2 & 3) would not show us a picture, but rather told us 'Not Found.' Also, answers were highlighted in a different color until we completed the questions. I'll talk to you again soon. Thanks again.

Here, Kathy expresses her excitement about the new technology she has learned and troubleshoots her lesson. Links need to be checked and additional "html" codes added to allow for the smooth operation of her lesson. Kathy, Nancy, and Karen have been exposed to a new avenue for teaching. Through the EARTH-WATCH-NYNEX grant they have been trained in field archaeology, collected firsthand information and skills, and learned a new language for sharing their ideas and lessons worldwide through the Internet.

Vince La Motta, one of the graduate students on site at *Homol'ovi I*, sheds light on the field of archaeology when he states: "Archeology is not a snapshot. It is a series of processes ... we try to unravel the mystery of these processes."

Figure 13. A *Homol'ovi I* field assistant uses her trowel to excavate an ash feature in the East Plaza area. (Photograph: Karen Travers)

This statement can be applied to the EARTHWATCH Education Program as well. We are a program that, through the support of donors such as Nynex, empowers teachers to take the controls to unravel the mysteries of field science, the intricacies of teaching, and the potential of technology. EARTHWATCH models process learning through firsthand experience.

EARTHWATCH EXPANDS LEARNING TO INFINITE REALMS

"The experiences that we have as teachers are multiplied many times as we use them to expand the world of our students." As underscored by Donna Davis,

Technology Coordinator at W. B. Patterson Elementary in Washington, D.C. and Bell Atlantic teacher fellow in 1996, experiential development and growth among teachers is the catalyst for exponential learning among students. Each teacher who returns from an EARTHWATCH project affects the lives and education of his or her students, thus multiplying the effect of worldwide scientific learning.

Liz Revenko brought back the skills of perseverance and field methodologies to her 100 Earth and Life Science students in California, Anne Steel and Samantha Sonnak connected their middle school students to firsthand radio tracking of black sea turtles, Cheri Ewald introduced over 100 fifth graders to the life of a research scientist, and Bob Frostick inspired his science students to head out behind their school to collect grasshoppers.

Each year EARTHWATCH projects and educational programs inspire over 200 teacher fellows to refresh their curriculum and emphasize hands-on science investigations. In turn, these teachers directly enlighten and educate more than 10,000 students in their classrooms nationwide. Through the latest technology of the World Wide Web, e-mail and Lotus Notes, EARTHWATCH expands its educational audience even further. No longer do the walls of a classroom or the property lines of a school limit learning. EARTHWATCH opens the door to infinite global learning and collaboration.

SOURCES

EARTHWATCH Materials

- EARTHWATCH Expeditions Inc. Web site (http://www.earthwatch.org)

    o Fields of Study (http://www.earthwatch.org/t/Tfieldsofstudy.html)
    o EARTHWATCH Global Classroom (http://www.earthwatch.org/ed/home.html)

- EARTHWATCH Expeditions Inc. Project Briefings.
- "EARTHWATCH: Bringing the Classroom into the Field." Posted on the National School Network. (http://nsn.bbn.com/community/earthwatch/earthwatch.html)

EARTHWATCH Teacher Reports

Greg Bisbee, *North America's Rarest Butterfly*, July 1996. Arrowhead High School, Hartland, WI. Biology, grades 9–11.

Donna Davis, *Forest Islands of Bolivia,* August 1996. W. B. Patterson Elementary School, Washington, DC. Technology Coordinator, grades K–6.

Bob Frostick, *Amazon Katydids,* August 1996. Horace Mann Junior High School, Charleston, WV. Science, grades 8–9.

Merey Grearson, *Black Sea Turtles of Baja,* June 1996. Tabor Academy, Marion, MA. English, grades 10–11.

Nancy Guenther, *Homol'ovi I,* July 1996. Braintree High School, Braintree, MA. History, grades 9–10.

Penny Jenkins, *Amazon Katydids,* August 1996. Western School of Technology & Environmental Science, Catonsville, MD. Science, grades 9–12.

Katherine O'Brien, *Early Roman Farm,* July 1996. Samuel Adams Elementary School, Boston, MA. All Subjects, grade 2.

Elizabeth Revenko, *Modeling Forest Canopies,* July 1996. The Bishop's School, La Jolla, CA. Science, grades 7–8.

Samantha Sonnak, *Black Sea Turtles of Baja,* July 1996. Springside School, Philadelphia, PA. Science, grades 6–9.

Anne Steel, *Black Sea Turtles of Baja,* July 1996. Oxford Middle School, Oxford, MS. Math, grade 8.

Karen Travers, *Homol'ovi I,* July 1996. South Intermediate School, Weymouth, MA, All subjects, grade 6.

Kathy Womersley, *Homol'ovi I,* July 1996. Gates Intermediate, Scituate, MA, Social studies, grades 6–8.

Other Sources

- Cheri Ewald, research scientist. Coordinator of Science Adventure Program, Matthews Elementary School, Charlotte, NC.
- Fifth grade students. Science Adventure Program, Matthews Elementary School, Charlotte, NC.
- Tally Forbes, Director of Education. EARTHWATCH Expeditions, Inc., Watertown, MA.
- Meg Warren, Curriculum Specialist. EARTHWATCH Expeditions, Inc., Watertown, MA.

Quoted EARTHWATCH Scientists

- Dr. Constance Becker. Principal investigator, *Community Forestry in Ecuador.* Assistant Professor, Indiana University. Director-Ecologist, People Allied for Nature, Ltd.
- Vince La Motta. Research Assistant, *Homol'ovi I.* Doctoral candidate, University of Arizona.

CHAPTER 7

Mars Exploration
Students and Scientists Working Together

Daniel Barstow and Cheick Diarra

In August 1996, NASA scientists made a stunning announcement. They found possible evidence of ancient life on Mars. These scientists speculate that the object shown in Fig. 1 might be a Martian microfossil, possibly the remains of primitive life from ancient Mars. Scientists used advanced technologies to examine a 4.5-billion-year-old rock that had blasted off Mars millions of years ago, traveled through space, and landed in Antarctica. This is the strongest evidence ever found of life on another planet.

This discovery raises profound scientific, philosophical, and educational questions. Are we alone in the universe? Was ancient Mars covered with water and teeming with early life? Is there still life on Mars? Where did the water on Mars go? How do we define "life"? These questions are at the heart of NASA's Mars Exploration Program and of an innovative collaboration between NASA scientists and students in schools throughout the country.

Prior to this announcement, NASA had already embarked on an ambitious 10-year series of robotic missions to Mars. The missions will conduct remote reconnaissance from orbit around Mars and on-site experiments and observations from the surface of Mars. The first two of these missions, an orbiter called Mars Global Surveyor and a lander called Mars Pathfinder, were launched in November and December 1996, respectively. As this chapter is being written,

Daniel Barstow, TERC, 2067 Massachusetts Avenue, Cambridge, MA 02140; banbarstow@terc.edu. Cheick Diarra, Jet Propulsion Laboratory, NASA, 4800 Oak Grove Drive, Pasadena, CA 91109; cheick.m.diarra@jpl.nasa.gov.

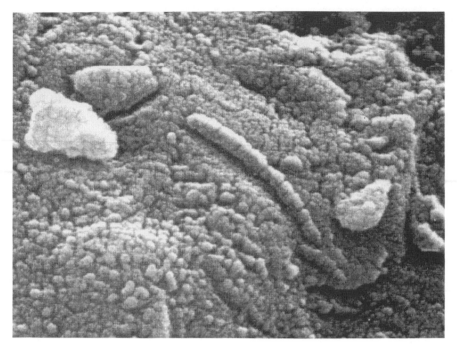

Figure 1. Possible Martian microfossil.

these missions are on their way to Mars. Other robotic missions will follow on a regular basis at least until the year 2003, and probably beyond.

As a result of the recently discovered evidence of possible ancient life on Mars, NASA convened an expert team of scientists to review the current and future Mars missions. The team is considering the evidence of life, trying to understand the implications, and developing proposals for modifying the goals and techniques for future missions to search more effectively for past and present life on Mars. This is a time for combining high-level scientific expertise with innovative thinking. It is in this context that NASA's Mars Exploration Program believes that a parallel process conducted by high school students working with topnotch planetary scientists will come up with innovative ideas that could contribute to meeting this challenge. Their recommendations will also include ways for students to be active participants in several aspects of the Mars missions.

This is potentially a powerful win–win situation. As described below, scientists gain student collaborators who could help with the planetary exploration and science research programs; and students gain an exciting context for learning science and its processes, by participating directly in authentic missions and research. The goal—the search for life on Mars—is arguably one of the most important science journeys of our time.

The Mars Exploration Program is managed at NASA's Jet Propulsion Lab (JPL) in Pasadena, California. JPL is the lead NASA center for planetary exploration and has managed planetary missions such as Voyager, Magellan, Viking, and the current Mars missions. Their educational partner in developing this student–scientist partnership is TERC, a nonprofit educational research and development company in Cambridge, Massachusetts.

Some of the efforts described will involve several hundred schools throughout the United States, at elementary, middle, and high school levels. The proposed "Search for Life" effort will involve a more limited group of about 100 students in four high schools. Most of the scientists involved work at JPL, with a few others at universities specializing in Mars and planetary research. Participating teachers will receive training and packets of support materials.

This chapter focuses on the ways in which students and scientists could work together in planetary exploration, the benefits for each side of this partnership, and the enabling technologies that support and foster this work.

GOALS FOR THE SCIENTISTS

We begin with a brief science background and a look at the goals and priorities for the scientists involved in this effort.

NASA's Mars Exploration Program involves a multifaceted, 10-year series of robotic missions to Mars. The spacecraft will conduct orbital surveillance and landed experiments. If successful, this will be the first return to Mars since the historic Viking missions of 1976. During the Viking missions, two orbiters provided scientists with photographic images of the entire planet, most with a resolution of about 250 m/pixel (Fig. 2). In addition, the Viking landers provided photos of the area around each of two landing sites, monitored local weather, and conducted biological experiments searching for evidence of life (Fig. 3). The results of that experiment are controversial, but are generally regarded as finding no evidence of life. Over the past 20 years, however, scientists have extended their studies of the images and data from Viking, and among other things, have learned a great deal more about Mars's geologic history, atmosphere, climatology, and hydrology.

Nevertheless, many questions remain. The focus of this new series of Mars missions, called the Mars Surveyor Missions, is on three broad domains:

1. Search for Evidence of Past or Present Life on Mars

From all appearances, Mars is a barren planet. There is no liquid water on the surface, the atmosphere is very thin (1/100th Earth's atmosphere), dangerous levels of ultraviolet radiation bombard the surface, and the surface is highly oxidized (the red color of Mars is essentially rust). Neither the Viking orbiter nor the lander cameras showed any signs of current life on Mars. However, the orbital images did reveal evidence of flowing water in Mars's ancient past. This

Figure 2. Mars hemisphere.

evidence has provided a context for thinking about the possibility that life arose
in that ancient past. The evidence of possible ancient life found in the Mars
meteorite lends strength to this speculation. And if there was life in the past, it
may have evolved to survive in pockets of subsurface water or elsewhere on or
in present-day Mars. The search for direct or indirect evidence of such past or
present life is a central research goal for Mars exploration missions.

2. Better Understanding of the Role of Water and the Hydrologic Systems of Mars

The search for life is closely tied to the search for water. Water is central
to life on Earth and is a reasonable starting point in the search for life on other

Figure 3. Viking I lander view of the Martian surface.

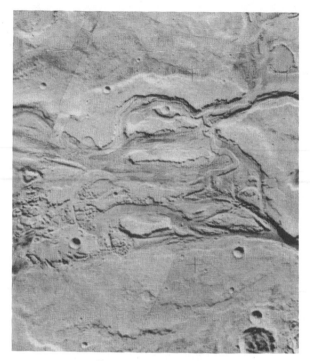

Figure 4. Martian channels.

planets. Even aside from the question of Martian life, learning more about the role of water in past or present Mars is pivotal to our understanding of the whole planet. The image in Fig. 4 shows what appears to be channels formed by ancient flowing water. Other images show what might have been lakes or oceans in the ancient past. Investigating Martian hydrology will include studying more detailed images from future orbital missions, data from other orbiting instruments, and surface measurements and experiments. Scientists want to better understand when and where water flowed in the past and whether there might still be subsurface liquid water and possibly life. This research domain also includes studying the weather and climatologic history of Mars.

3. Identifying Resources on Mars to Support Further Explorations

Future missions to Mars, possibly including human missions, will probably need to take advantage of resources existing on Mars. For example, rather than transporting fuel to Mars, fuel could be manufactured on Mars by creating combustible methane using carbon dioxide in the Martian atmosphere and liquid hydrogen brought from Earth. There are likely to be mineral ores and other useful materials on or just below the surface. There may even be liquid water at deeper levels. Aside from the potential for supporting future missions, better understanding of the resources and composition of the Martian surface is of prime scientific importance.

GOALS FOR THE STUDENTS

Recognizing the need to improve science education in this country, the National Research Council of the National Academy of Sciences conducted a broad review of science education in the United States. and made a set of recommendations. These recommendations, called the Standards, define criteria for assessing and improving the quality of science education programs. The recommendations deal not only with the science content, but also with the science thinking skills developed in the students, and the overall approaches to science and teaching and learning. The *National Science Education Standards* have challenged schools to transform science education and proposed criteria for guiding and assessing this transformation.

Among other things, the Standards call for:

- *Inquiry-based learning*—student investigations should arise from their own questions and they should use direct observations, experiments, and primary sources of data to figure out the answers.
- *Engaging students in real science*—students should participate in the process of science, in the context of real scientific investigations.
- *Using scientific thinking and problem-solving*—students need to experience the rigor of scientific thinking and experimentation; when problems arise, students need to develop their own solutions
- *Increasing use of technology*—students should use computer, telecommunications, and other technologies, to help them pursue their investigations.
- *Integrating among the science disciplines*—students need to understand that the various domains of science are all interconnected, and that one cannot do research in one domain without knowing about other related domains.

Let's consider these criteria in the context of learning about Mars. Most students learn about Mars by reading a science textbook. They see a couple of pictures of Mars and perhaps a graph, and then read what scientists have discovered. Then they answer a few questions at the end of the chapter on "planets." This is neither good science nor good education. Students need to participate in the process and thrill of scientific discovery, they need to learn science by doing science. This is the core challenge affecting science education today.

The Mars Exploration and Education Program is one attempt at such a solution. By providing opportunities for students to participate in real planetary research and work directly with NASA scientists, students will:

- *Do inquiry-based learning*, by asking questions and finding answers using primary data
- *Engage in real science*, by exploring questions at the cutting edge of science; use scientific thinking and problem-solving, by working with scientists as they plan for, conduct, and use data from robotic missions to Mars
- *Use technology* to access data and images in real time, and telecommunicate with other students and scientists
- *Integrate among the science disciplines*, as they explore issues that combine astronomy, geology, meteorology, biology, hydrology, and comparative planetology

THREE EXAMPLES OF THE STUDENT–SCIENTIST COLLABORATION

The power of student–scientist partnerships is that, if done well, they can meet the needs of both the students and the scientists. This overlap can be illustrated by three examples. These examples are vignettes for the future, as the first of the new Mars missions will not arrive at Mars until July 1997. In anticipation, NASA JPL, TERC, and participating teachers and students are currently defining, recommending, and prioritizing the domains of potential collaboration. The examples described below are potential elements of this collaboration.

As a key starting point, a set of Mars education modules has been developed to introduce students to Mars, planetary science, the process of image analysis, and planetary exploration. These modules use a hands-on approach, employing classroom experiments and data and images from previous Mars missions to support student investigations. Students in hundreds of schools have already begun using these modules, in preparation for the more in-depth research partnerships described here.

Example 1—Studying Images

The Mars Global Surveyor will orbit Mars for at least a full Martian year (687 days). As it orbits, it will use a variety of instruments to study Mars. Most notable are the cameras that will take wide-angle and closeup pictures of Mars, day after day after day. They will produce images faster than scientists can review and catalog them. When properly trained, students can provide a first pass review of selected images, looking for specific features, such as evidence of dust storms or valley fog. Because such dust storms and fog are short-duration events, students finding such features will quickly notify the scientists, who can then

Figure 5. Closeup of Valles Marineris.

launch more detailed examinations. Students also extend their own learning by examining the same images in more detail on their own.

Scientists gain an "early warning system," and the students gain an entry point into the study of current images from Mars.

Example 2—Explaining Mars Research to the Public

Every week, the Mars Exploration Program will release a new set of images to the public, available on JPL's Web site. However, it is not enough simply to post the images. Interested people need interesting questions and some guidance to help them engage in thinking about the images. Students who have experience in image analysis, understand key concepts in Mars science, and know how to communicate concepts clearly, can serve as "translators" from the technical science to the language and concepts the general public might understand. They also can suggest "mini-investigations," such as crater counts or comparing different types of channels. Students can also serve as spokespeople for local newspapers or TV stations. Such students will need to be "certified" to ensure the accuracy of the information they have communicated.

Scientists gain improved public support and involvement in the Mars missions, and the students develop science content knowledge and communication skills.

Example 3—Aiding in the Search for Life on Mars

The most advanced proposed collaboration involves teams of high school students and a select group of NASA scientists involved with the Mars Exploration Program. The students will work directly with the scientists to understand the challenges of the search for evidence of past or present life on Mars, develop recommendations for strategies to pursue this search, and propose real and substantive ways that students can be active collaborators in this scientific research over the next decade. The project will operate as a national network. Four regional teams will be identified, each with a scientist, teacher, and students. Each team will have a focal domain: exobiology, geology, climatology, and engineering. The focal domains enable the scientists and students to work intensively within each domain, whereas the collaboration enables the work of the teams to be coordinated and to proceed toward a cohesive set of recommendations. The recommendations will be reviewed and seriously considered by JPL's Mars Exploration Program.

Scientists gain a fresh new perspective and potentially innovative ideas which can help improve the Mars Exploration Program. Students gain the opportunity to work directly with some of the world's leading planetary scientists, have access to the latest data and images from Mars, and have the responsibility for preparing recommendations to NASA.

MAKING THE PARTNERSHIP WORK

There are three key elements in this partnership which must be emphasized:

1. *Scientists and students must each have real personal motivations.* For the scientists, this is not just helping improve education, it is helping the scientists with their own work in real and important ways. For the students, this is not just being "cheap labor," they must have a genuine interest in planetary exploration. Space exploration has a natural appeal so this shouldn't be too difficult.
2. *The collaboration must involve authentic science research.* The students should not be doing artificially crafted "research," they should be meaningfully engaged in the science research process. For most

Figure 6. Authentic science and education partnerships.

students, this will be a totally new and potentially very exciting aspect of their science education experience.

3. *The collaboration is not the total experience for either party.* Although the partnership must involve an overlap between science and education (as illustrated in Fig. 6), the parts of the circles that do not overlap are just as important as the parts that do. For example, when students serve as the "early warning system" spotting dust storms or valley fog for the scientists, the students' learning experience should go far beyond that experience, by learning about Martian atmosphere, image analysis skills, the implications of dust storms, comparisons of Earth and Mars, and so on. In other words, for the students, the collaboration is an exciting "hook" that engages them in real science research, but the classroom experience serves as a springboard to other learning activities to deepen and extend the experience. Similarly, the scientists will use their own advanced knowledge and research skills in ways that entry-level students should not be expected to understand. For the scientists, the help from the students is likely to be just one small, albeit important, part of their full scientific research activities. The collaboration is the intersection of the work of the scientists and the learning of the students, but is not the totality of either's experience.

THE ROLES OF TECHNOLOGY

Technology is central to our ability to achieve the science and education goals of the Mars Exploration and Education Program. In fact, robotic planetary exploration uses some of the most advanced technology in the world (or, more correctly, in the solar system). When a mission is on its way, it needs to be both solid and flexible. No one can make a service call to the depths of space, so the hardware must be completely reliable. On the other hand, the mission must be modifiable, with the spacecraft accepting new commands to fine-tune the flight path and the use of the instruments in response to changing circumstances or any technical problems that might arise. These are exceedingly challenging demands on the technology.

Terrestrially, advanced technology is also central to the communication, display, analysis and sharing of the data and images. The scientists and students using the data essentially can be anywhere in the world, as long as they have sufficient bandwidth for telecommunications or access to high capacity data storage devices. NASA has made a major commitment to public distribution of data and images from the Mars mission via the Internet and the World Wide Web. This policy and technology enable students in the Mars Exploration and Education Program to have direct access to the same images and data as the scientists, or at least a sizable subset thereof.

More specifically, there are four amazing technologies supporting the Mars Exploration Program, as follows.

1. Robotic Technology of the Missions

There are two missions currently en route to Mars, as of the publication date of this chapter.

Mars Pathfinder is a lander, with a remarkable landing technology (Fig. 7). It was launched on December 4, 1996, and is scheduled to arrive at Mars on July

Figure 7. Mars Pathfinder.

4, 1997. Mars Pathfinder's descent will be slowed by a heat shield, parachutes, and retrorockets. Then, balloons will inflate around the spacecraft and it will literally bounce across the surface until it comes to a stop. The balloons will deflate and petals of the spacecraft will unfold like a flower, revealing the instruments and a rover.

The instruments include a camera, which will provide stereo, panoramic views of the area around the landing site, and weather instruments to monitor temperature, wind speed, wind direction and atmospheric pressure. Students will be able to access the camera's images and graph data from the weather sensors.

Mars Pathfinder also includes a mobile rover, called Sojourner. It was named by a middle school student, after the African-American abolitionist Sojourner Truth. The rover, which is about the size of a microwave oven, will roll off the lander and explore the area within roughly 100 m of the lander. It is semiautonomous, in that a person on Earth will periodically guide it in a specific direction, and then Sojourner will proceed, using a laser guidance system to go over or around rocks and pits in its way. Sojourner has a camera, so we will see the Martian landscape as the rover traverses the surface. It also has a device called an Alpha-Proton X-Ray Spectrometer which can determine the elements in the rocks and soil it samples.

Sojourner will explore for an expected minimum of seven Martian days to a hoped for maximum of 30 Martian days. The primary limiting factor is the solar panel, which might get covered by Martian dust. A group of students suggested a windshield wiper-like device to remove accumulated dust; however, it was too late to change the design.

Mars Pathfinder will land in the Ares Vallis region of Mars (Fig. 8). This is an especially interesting area, in that it is believed to be the primary drainage area of an ancient river channel. The flooding that created the channel about 3 billion years ago is likely to have carried many different types of rocks to the region of the landing site. Thus, Mars Pathfinder will help scientists and students understand the geology and hydrology of both the landing site itself and the regions through which the water flowed. It will be quite exciting to explore the surface of Mars with the wandering Sojourner!

Mars Global Surveyor is an orbiter that will use remote sensing to explore Mars from a height of about 380 km (Fig. 9). Launched November 7, 1996, it will arrive at Mars in September 1997, take 6 months to aerobrake and stabilize its orbit, and then begin a reconnaissance mission lasting at least a full Martian year (687 Earth days).

Mars Global Surveyor has several instruments. Of primary interest in the student–scientist collaboration is the Mars Orbital Camera. Actually there are two cameras: a wide-angle camera which will provide a daily overview of the entire planet (useful for monitoring weather and dust storms), and a telephoto camera which can show detailed closeup views of selected areas of Mars (1.4 m/pixel). The closeup views will enable scientists and students to get detailed views of craters, canyons, volcanoes, poles, and other important features of Mars.

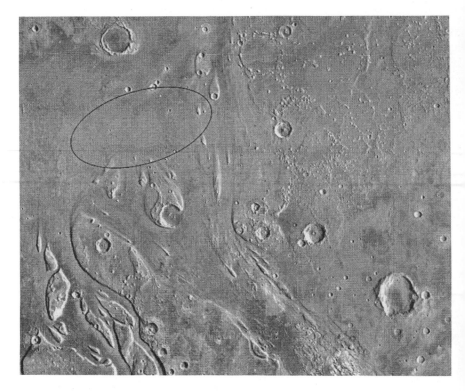

Figure 8. Mars Pathfinder landing site.

Other instruments on Mars Global Surveyor include the thermal emission spectrometer (to measure heat emitted from the planet), the laser altimeter (to measure height of Martian surface features) and the magnetometer and electron reflectometer (to study Mars's magnetic fields). Collectively these instruments will map the planet in new ways, provide rich resources for investigations, and potentially reveal new discoveries that will change our understanding of Mars.

As mentioned earlier, we are defining ways for students to work with scientists in the analysis of the wealth of images, especially from the Mars Orbital Camera. In some cases, students might help in the initial review of images, looking for signs of dust storms or valley fog, or further evidence of ancient flowing water. In other cases, students might do preliminary crater counts and analyses (the size and nature of craters help determine the age of a region). It is also possible that more advanced students will be able to select specific surface targets for the Mars Orbital Camera to photograph as the students pursue their own investigations.

Future missions. Other robotic missions will be sent to Mars over the next decade. Every 26 months, the planets are aligned for the most energy-efficient path

Figure 9. Mars Global Surveyor.

to Mars. The next launch opportunity is January 1999, at which time another orbiter and lander will be sent to Mars. Each mission has unique objectives and instruments. The next lander will land close to the Martian south polar region, and will be able to dig down into the soil, investigating soil composition and searching for more information about water beneath the surface. In approximately 2003, a robotic sample return mission is planned, bringing rocks back to Earth. We are also exploring ways for students and scientists to work together on these future missions.

2. Telecommunications Access to Data and Images

The Internet and the World Wide Web are transforming space exploration from the exclusive domain of a select group of scientists to a set of resources that are widely available to the general public. In past missions, NASA selected, printed and widely distributed a handful of dramatic images for each mission. Now, for this new generation of Mars missions, NASA has made a major commitment to posting a large number of images and data from the Mars missions on the World Wide Web, with a new set of images posted every week (Fig. 10).

This expansion in the distribution policy means that the general public, including students, can see the images virtually in real time, and can peruse a wide selection of images to pursue investigations of their own interest. Data from the Pathfinder's weather sensors will also be posted, so that students can monitor, for example, daily temperature changes on Mars and compare Martian temperature cycles with their own local temperature cycles.

Real-time image distribution is essential for the student–scientist collaboration. Without the World Wide Web, this collaboration could not take place, or would be limited to schools that are located close enough to frequently visit the JPL facilities in Pasadena, California.

3. Visualization Tools for Image Analysis

Dramatic advances have taken place over the last decade in the tools of image and data visualization. Although many of the collaborative investigations,

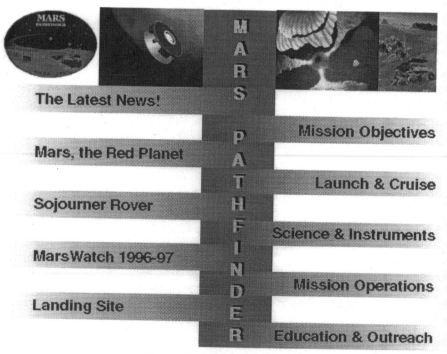

Figure 10. JPL's Mars Web site.

such as crater counts, involving image analysis can be done with simple display tools, advanced visualizations can extend the investigations further.

For example, the image in Fig. 11 shows a surface view of Venus. This image was created entirely by computer-based image manipulation. The source of the data is the Magellan radar, which orbited Venus from 1990 to 1993, collecting detailed elevation data for the entire planet. Computer visualization tools were then used to create this simulated surface view. No camera has ever seen this view—it is a creation of the visualization tools, and is essential for the scientists to do their research. Video fly-over sequences are also used to help scientists visualize and understand the data. The IMAX movie "Destiny In Space" includes a dramatic video sequence of a simulated Venus flyover.

Such visualization tools will be especially important for the Mars Global Surveyor's laser altimeter. The data from the altimeter will simply be a set of data points indicating latitude, longitude, and elevation. It will require visualization tools to transform these data into images and/or video sequences. The visualizations will be highly manipulable, enabling scientists and students to create views of the surface from a variety of different perspectives.

Figure 11. The surface of Venus.

It is important to point out that although the computer-generated images provide a rich resource for data visualization and analysis, the resulting images often use false colors and vertical exaggerations. This enhances their power for research, but can lead to misperceptions, especially for a naive lay audience. For example, the image of Venus shown in Fig. 11 has a 10× vertical exaggeration (the mountain appears 10 times higher than it is). In reality, the surface of Venus is much flatter than it appears here. This image in particular has been widely distributed and has created a popular, yet false, view that Venus actually looks like this.

Some of the students in the Mars Exploration and Education Program will experiment with visualization tools. Although the most powerful tools require supercomputers, there is an increasing set of graphics tools that can be used on personal computers. For example, Mars data can be used with some of the Geographic Information Systems (GIS) software now available for personal computers.

One of the benefits of the student–scientist collaboration in this area is that students can help test and propose refinements to some of these tools, especially in terms of the user interface. Given the extensive experience many students have with video games, which often include geographic visualizations, flyovers, oblique threee-dimensional perspectives, and so forth, students will very likely have valuable insights to make the scientists' visualization tools more accessible to the general public.

4. Tools for Telecollaborations

The fourth category of enabling technology is "tools for telecollaborations." As detailed earlier, select groups of high school students will work directly with planetary scientists to review the Mars Exploration Program, the scientific research priorities for the search for life on Mars, and the spacecraft designs and instruments available over the next decade to support this research. The students and scientists will work in four teams (exobiology, geology, climatology and engineering) to do their research and analysis, and then make recommendations to JPL for potential ways to enhance the search for life.

This work will require a considerable degree of communication and collaboration. On one level, the students on each team will communicate regularly with their scientist colleagues. On another level, the four teams will communicate and collaborate among themselves. Finally, the students will communicate their findings to JPL. All of these efforts require tools of telecollaboration. As the collaborators don't share a common physical space, the best solution is to share a common cyberspace.

The three primary communication tools are e-mail, telephone, and World Wide Web.

Each participant will have an e-mail account, and e-mail messages will be the most common way to communicate. E-mail is virtually instantaneous, enables asynchronous communication, and can be used to communicate the same message(s) to many people simultaneously. While telephones are so common that we usually do not think of them as technology, monthly telephone teleconferences will support dialogue and synchronous conversations.

The third tool is the World Wide Web. In addition to providing access to the Web's vast resources, including JPL's Mars Exploration home page, the Web is a marvelous workspace for collaborative development among the teams. Each team will have its own Web page, on which each team will post text and graphics to convey key concepts in their respective domains, the evolving plans and recommendations, and the linkages with the work of the other teams. The Web will also provide a common workspace for sharing ideas, images, drafts, and so forth. Documents and reports will be prepared and collaboratively reviewed on the Web, to enhance the words and illustrations, find and fix errors, and build a consensus on the full reports and recommendations.

A GRAND EXPERIMENT

This is an experiment. Never before have precollege students and scientists worked together in such a multifaceted collaboration in planetary exploration. We don't yet know how effective this will be in meeting the goals of the scientists or the goals of the students.

However, there are enough signs of the potential for both sides to gain for us to move forward in the planning, implementation, and careful review of this innovative and ambitious collaboration.

The educators believe that the students will gain an exciting and powerful context and stimulus for learning science content and for experiencing the real process of science. This science education program can be a stellar example of the reforms called for by the National Science Education Standards: inquiry-based learning, engaging students in real science, using scientific thinking and problem-solving, increased use of technology, and integrating among the science disciplines.

The scientists believe that educational outreach is an essential component of the space exploration program, and that this program provides a very interesting new approach to this outreach. Further, they are genuinely intrigued by the potential for this collaboration to help them with their own research in the primary research domains of the Mars Exploration Program: search for evidence of past or present life on Mars, better understanding of the role of water and the hydrologic systems of Mars, and identifying resources on Mars to support further explorations.

Figure 12. Scientists exploring planetary images.

Figure 13. Students exploring planetary images.

Perhaps the most important gain will be the realization by the students and scientists that the work they each do is not all that different. Both scientists and students do their best work when they are actively engaged in hands-on investigations of real-world problems of personal interest.

CONTACTS

To participate in the Mars Exploration and Education Program, and/or to obtain copies of Mars education materials, contact Dr. Cheick Diarra, Mars Exploration Education and Public Outreach Program, Jet Propulsion Lab, 4800 Oak Grove Drive, Pasadena, CA 91109, 818-354-6111.

To monitor the progress and obtain images and data from the Mars missions, use the Web: http://www.jpl.nasa.gov/mars/

CHAPTER 8

Validating Student Data for Scientific Use
An Example from the GLOBE Program

Russell G. Congalton and Mimi L. Becker

INTRODUCTION

The need for accurate data is universal. For scientific use of data collected by students, high accuracy is a prerequisite. Valid student-collected data for scientific purposes is of great benefit, both to the student who experiences science firsthand and to scientists who then have a larger database of information. Therefore, projects and techniques that provide for validating student data are extremely useful. Such a technique was applied in a program called GLOBE.

Global Learning and Observations to Benefit the Environment (GLOBE) is a hands-on international earth systems science research and environmental science education program that by design tightly couples science and education (see Chapter 2). The GLOBE program is intended to foster the creation of a worldwide research team, comprised of 5- to 18-year-old students and their teachers in collaboration with environmental scientists, for the purpose of generating knowledge about the Earth as an interconnected system. Presently, this team collaborates to facilitate both the learning and doing of science by students from over 3000 schools in 42 countries worldwide.

GLOBE students become the "environmental experts" for the 15 by 15-km study sites that surround their schools. From these areas, they collect specified measurements and transmit them through the GLOBE home page on the World

Russell G. Congalton and Mimi L. Becker, Department of Natural Resources, 215 James Hall, University of New Hampshire, Durham, NH 03824.

Wide Web to the GLOBE archive computer at NOAA in Boulder, Colorado. The World Wide Web also facilitates communication between research scientists and students. GLOBE learning activities have been designed to promote students' understanding of science through the use of tools such as global positioning systems (GPS), visualizations, and satellite image analysis. Students collect and report specific information about the atmosphere, hydrology, soil, land cover, and biology of their local GLOBE sites, through the use of standardized instruments and protocols. The science teams have designed environmental measurement protocols, appropriate for use by primary and secondary school students, and intended to ensure accurate and reliable measurement and reporting of data for use by the international environmental science community. Research teams at the University of New Hampshire have developed the learning activities as well as protocols necessary for collecting land-cover and forest/grassland biometry data used in validating land-cover maps generated from remote sensing. These activities have been evaluated by teacher focus groups and extensively tested by students in the field. Now they are being implemented at GLOBE schools.

Collecting accurate ground validation/reference data is fundamental to the use of remotely sensed data for land-cover classification and mapping. Such validation/reference data are necessary for the digital classification process and to assess the accuracy of results. The accuracy of land-cover maps is particularly important when such data are to be incorporated as a layer in a geographic information system (GIS). It is now common knowledge that error propagation through the layers of a GIS can be substantial. However, little ground valida-tion/reference data have actually been collected for most large-scale projects that utilize satellite imagery. A large amount of data are required if we are to effectively monitor the state of the environment on a continental to global level. GLOBE provides an excellent opportunity for participating students to collect much needed, large quantities of ground validation/reference data. However, a key question is, can the student-collected data provide an accurate enough basis to support rigorous scientific investigations?

BACKGROUND

The technique used in the GLOBE program to validate maps derived from satellite imagery is called an error matrix (Congalton, 1991). A statistician would call this same matrix a contingency table (Bishop et al., 1975). In either case, this error matrix/contingency table is simply a cross-classified table of counts, in which the sample units are cross-listed according to several categorical variables (Fienberg, 1980). For example, we may want to tabulate the results of an experiment in which some volunteers get a flu shot and others do not and cross-classify that with who got the flu that year (Table 1). The hypothesis tests if the flu shot reduces the risk of getting the flu looking at four possible pairs of

Table 1. An Example of a
Generic Contingency Table/Error
Matrix

	No Flu	Got Flu
Had Flu Shot		
No Flu Shot		

conditions: (1) had flu shot and did not get the flu, (2) had flu shot but got the flu, (3) did not get flu shot and did not get the flu, and (4) did not get flu shot and got the flu.

In validating student data, the cross-classification of interest compares the student answers with a set of validation/reference data. This matrix approach is applicable for many different kinds of categorical data. For instance, we may want to test student ability to determine cloud type. A matrix can be generated with student answers on the y axis and the correct answers (validation/reference data) on the x axis (Table 2). Not only does the matrix allow the students to see which clouds they correctly identified, but it also allows them to see which clouds they are confusing with one another. Such a matrix could also be generated for bird identification (see a complete example in the Learning Activity at the end of this chapter), for dominant land-cover type for a specific area, and for many other examples.

Table 2. An Example Error Matrix for
Evaluating the Accuracy of Student
Cloud Identification

REFERENCE DATA

STUDENT ANSWERS	Cumulus	Stratus	Cirrus
Cumulus			
Stratus			
Cirrus			

In the GLOBE program, we used an error matrix to assess (i.e., validate) the accuracy of land-cover maps derived from satellite imagery. Like Tables 1 and 2, a remote sensing error matrix is simply a square array of numbers set out in rows and columns, which express the number of sample units (i.e., pixels, clusters of pixels, or polygons) assigned to a particular land-cover category, relative to the actual category as verified on the ground (Table 3). The columns usually represent the validation/reference data and the rows the classification generated from the remotely sensed data. An error matrix is a very effective way to represent accuracy in that the accuracies of each category are plainly described along with both the errors of inclusion (commission errors) and errors of exclusion (omission errors) present in the classification. A quick look at Table 3 demonstrates these concepts. The value of 65 in the cell along the major diagonal indicates that 65 times the student classification said the area was forest and the reference data agreed that it was forest. However, 9 times the student classification said it was agriculture when the reference data said it was forest. In other words, 9 times an area was misclassified as agriculture when it was really forest. This means that 9 areas were committed to agriculture and omitted from the forest.

Next, the error matrix can be used as a starting point for a series of descriptive and analytical statistical techniques (Congalton, 1991). Perhaps the simplest descriptive statistic is overall accuracy, which is computed by dividing the total correct (i.e., the sum of the major diagonal) by the total number of samples in the error matrix. Accuracies of individual categories can be computed in a similar manner.

Table 3. An Example Remote Sensing Error Matrix

Reference Data

		F	A	W	D	row total	
Classified Data	F	65	7	2	0	74	Land Cover Categories
	A	9	81	1	3	94	F = forest
	W	0	1	85	0	86	A = agriculture
	D	0	2	3	90	95	W = water
column total		74	91	91	93	349	D = Developed

OVERALL ACCURACY = 321/349 = 92%

PRODUCER'S ACCURACY

F = 65/74 = 88%
A = 81/91 = 89%
W = 85/91 = 93%
D = 90/93 = 97%

USER'S ACCURACY

F = 65/74 = 88%
A = 81/94 = 86%
W = 85/86 = 99%
D = 90/95 = 95%

However, this case is a little more complex, the number of correct pixels in any category can be divided by either the total number of pixels in the corresponding row or the corresponding column. Traditionally, the total number of correct pixels in a category is divided by the total number of pixels of that category as derived from the validation/reference data (i.e., the column total). Such an accuracy measure indicates the probability of a validation/reference pixel being correctly classified and is really a measure of omission error. This accuracy measure is often called "producer's accuracy" because the producer of the classification is interested in how well a certain area can be classified. On the other hand, if the total number of correct pixels in a category is divided by the total number of pixels that were classified in that category, then this result is a measure of commission error. This measure, called "user's accuracy" or reliability, is indicative of the probability that a pixel classified on the map/image actually represents that category on the ground (Story and Congalton, 1986).

PROTOCOLS AND ACTIVITIES

The GLOBE land-cover team developed scientific protocols and supporting educational activities to prepare elementary, middle, and high school students to accurately collect land-cover data. These data are used by scientists and students to generate and evaluate maps created from satellite imagery. The scientific protocols instruct students how to establish qualitative and quantitative land-cover accuracy assessment sites. At forested sites, students identify the dominant and codominant tree species, measure tree height and circumference, measure the percent canopy and ground cover, and assign an appropriate land-cover class to the site. At grassland sites, appropriate grassland biometry is performed. Students use satellite imagery, image processing computer software called *MultiSpec* (Landgrebe and Biehl, 1995), and the ground-collected land-cover data to generate a land-cover map of the area surrounding their school. The map can be generated manually or by using image processing software on the computer. Once the map is completed, students can assess the accuracy of the map, using a combination of qualitative and quantitative methods. At a qualitative validation site, students determine the land-cover class using satellite imagery and ground observations. At quantitative accuracy assessment sites, students use more detailed ground measurements to assign a land-cover class and compare that to the land-cover class from the satellite imagery. All site locations are determined using GPS, allowing the land-cover maps to be referenced to coordinates in a GIS. Ultimately, the land-cover maps generated by GLOBE students, in conjunction with the site-specific information gathered about the atmosphere, hydrology, soil, and biology, can be analyzed in a GIS that will facilitate large-scale environmental monitoring.

A series of preprotocol educational activities were developed to provide students with background information on remote sensing, the collection and use of land-cover data for thematic mapping, and accuracy assessment. Postprotocol activities were developed to provide students with the opportunity to interpret and synthesize the data collected for the scientific protocols. The educational activities were developed, revised, and thoroughly tested by a focus group of teachers from southern New Hampshire and Maine. The focus group evaluated the appropriateness of the activities for specific age levels, clarified instructions and diagrams, and improved procedures for practical application in the classroom. In addition, each member of the focus group field tested the activities in his or her classroom and suggested revisions based on feedback from the students.

Currently, the accuracy assessment protocols are being tested in the field with a pilot group of volunteer teachers, camp counselors, and students. The teachers and camp counselors attend a 1-day workshop to learn the data collection protocols and activities. On completing the workshop, the teachers and camp counselors train their students to make the necessary ground measurements. In addition, students complete activities that provide background information on remote sensing and accuracy assessment concepts. After completing the training and practice activities, each group travels to our pilot testing site at Pawtuckaway State Park, New Hampshire to complete the fieldwork at a designated study site where we have a set of reference/validation data to compare to the student data. All of these results are compiled and sent to participants for follow-up discussion and analyzed in detail by the Land-cover team.

ADDITIONAL ONGOING RESEARCH

In addition to the development and creation of protocols and learning activities, our GLOBE research project has two other very useful components. The first is to evaluate the accuracy of land-cover data collected by GLOBE students. Student-collected data will be compared to a validation/reference data set generated by scientists (i.e., professionals) following the same data collection protocol used by the students. Participating students fill out a written evaluation of the training and the fieldwork to assess their skill level and their understanding of the project's objectives. Teachers provide an evaluation of the activities successfully completed in the classroom to document the level of training received by each student group. With the data from the fieldwork and the evaluations, we will determine the overall accuracy of the data collected by students using the error matrix approach. We will then determine whether errors in land-cover site assessment are correlated with the student's age, level of training in the research protocols, or understanding of the project's objectives.

Student data accuracy may be affected by difficult site conditions. Bad weather, bothersome insects, obstacles, or uneven terrain make site evaluation more challenging. As all of the data collection points are located by using a compass and pacing, we anticipate that students may have more difficulty staying on the proper compass bearings at sites with large obstacles and uneven terrain. Students also may have difficulty distinguishing which trees are part of the overstory at sites with a very dense understory; this increases the potential for errors in land-cover classification.

Land-cover classification will be more accurate when students are given sufficient training and time to practice. Each teacher must document the activities that were altered or eliminated, and the time spent on the completed activities, as well as the activities or concepts that were difficult for the students to comprehend. We expect that the student groups who have completed and understood all of the recommended activities will provide more accurate land-cover data.

Finally, we expect that the students will be more accurate when they clearly understand the objectives of this project and the overall objectives of GLOBE. We believe students who understand the importance of their scientific contribution will strive to do their best work. The results of the student evaluations will document whether they have gained this understanding from the preprotocol activities.

Provided that the student data are found to be accurate (i.e., at least comparable to that collected by professionals), this research will serve to increase acceptance of this method of data collection within the scientific community, and will enhance science education by providing students with a unique "hands-on" research experience. Accurate student data from 43 countries currently participating in GLOBE will begin to build the foundation of a truly global database necessary for environmental monitoring.

Another associated ongoing research project explores the specific relationship between the level of effort expended to collect land-cover ground validation/reference data and the associated accuracy of the data. If we are to effectively monitor the global environment and to identify the occurrence of changes, we must gather far more ground validation/reference data than is commonly done; thus, it is crucial to know how much effort is required to collect sufficiently accurate ground data. This question is relevant for both student- and professional-collected data.

When a satellite image is classified (i.e., labeled) into a thematic map of land-cover features, it should always be accompanied by a rigorous accuracy assessment. To perform a valid accuracy assessment of a classified satellite image, one must collect ground validation/reference data. There is always the question of, "How much effort must we expend to collect data that are 'correct'?" In this project, we are compiling data sets to compare qualitative ground observations of land-cover characteristics with quantitative measures of the same areas. (For this project, we define a "qualitative observation" as one in which no measurements are performed, only observations are made.) Clearly, the quantitative measurements require more time—we will quantify the relationship between this added effort and the accuracy of the data.

Procedures developed to test this relationship include multiple data collection protocols, with several intensities of effort within each of the quantitative and qualitative data sets (i.e., low-intensity qualitative, high-intensity qualitative, low-intensity quantitative, and high-intensity quantitative). Thus, the data include a broad spectrum of levels of effort for data collection. It is our hypothesis that qualitative measurements may be adequate to determine species composition. However, we anticipate that fairly intensive measurements will be necessary to collect canopy cover measurements of sufficient accuracy. Canopy cover is particularly important for many ecosystem models now being used and developed in conjunction with GIS databases. We expect to find that the more detailed level of classification will require a more intensive ground data collection effort to get accurate data. The utility of the different data collection methods may also be affected by the homogeneity of the tree species distribution within the forest stands. We will test the hypotheses that evenly dispersed (homogeneous) stands may be measured most accurately with ground measurements, whereas uneven or patchy stands may be more accurately represented by qualitative observation.

These data sets will also be compared to the ground validation/reference data collected by GLOBE students using the error matrix technique. The GLOBE students will collect field data in the same forest stands in which our other data are being generated. They will collect qualitative data and two intensities of quantitative data; assessing canopy composition, canopy closure, and ground cover. Our results will demonstrate whether GLOBE student-collected data are sufficiently accurate to meet professional scientific standards. They will also demonstrate what intensity of ground data collection is required of the students to achieve the desired accuracy.

Introductory Accuracy Assessment Activity

In fulfilling both its science research and educational objectives, GLOBE is designed to ensure that educators and students are clear about its scientific objectives and have developed understanding of scientific concepts and the skills that are required to do the investigations. Thus, students are not being taught about the scientific method just from textbooks. Instead they are engaged in experiential learning activities that equip them to collaborate with scientists in real, ongoing investigations to better understand earth systems. Additionally, GLOBE provides students and teachers with the opportunity to investigate earth systems problems of special interest in collaboration with other GLOBE schools across five continents. As GLOBE and other earth system monitoring programs such as Forest Watch, have been developed, it has become obvious that for scientists to use student collected data, they must have confidence in the data's accuracy. As our GLOBE investigative team proceeded to develop the Data Validation and Accuracy Assessment Protocols, it was clear that we needed to develop adequate conceptual understanding of our approach to accuracy assessment as well as to ensure that students developed classification and data assessment skills necessary for developing and interpreting an error matrix. A series of learning activities were designed, field tested with teachers and students, and adapted for this purpose.

An "introductory" or "preprotocol" learning activity targeted to primary or "beginning" participants in the GLOBE program is presented below to illustrate how such learning activities can be used to facilitate student understanding and skill building to aid the evolution of their ability to undertake the scientific protocol. In designing the learning activity demonstrated below, we used a classification system with a single decision rule—the shape of a bird's beak—to determine whether the bird was a carnivore, herbivore or omnivore. A more advanced learning activity introduced more experienced or advanced students to classification of cloud types, where the decision rules are more complex. This activity not only assists them with development of their accuracy assessment skills which will be applied to the GLOBE land-cover investigation, but also provides a basis for more accurate reporting of weather data which are required under a separate GLOBE investigation.

In presenting the learning activities we identified the student skill level required. In the case of the Bird Classification exercise this is at an "introductory" level. Young elementary students can be engaged to improve their science and math skills with this activity. A specified prerequisite is the ability to do basic classification. Students were also expected to have a baseline acquaintance with fractions and percentages (although the error matrix can be built from Leggo blocks rather than being displayed as a calculation). Recognizing that classroom time is limited, the activity is intended to be completed within one class period (45–50 minutes). The learning activity is intended to help develop students': (1) conceptual understanding of a scientific classification system; (2) ability to use criteria to guide decision making; (3) ability to evaluate the accuracy of their own work; (4) ability to improve the accuracy of their work; and (5) ability to analyze data. They also learn the importance of precision and how to use validation data. Specific skills to be developed include those of building an error/difference matrix, using an error matrix to improve accuracy, and cooperative problem solving.

As an introduction to the learning activity, the "global connections" to GLOBE protocols and the big picture were stated in the teacher's section of the briefing:

> Scientists classify many features of our environment such as species of life, forest types or soil types. These classifications are a fundamental mechanism for helping us to organize and to understand the natural world. There may be several different appropriate ways to classify a set of objects of interest. Two particular objects may be classified differently either because of error on the part of one or both of the classifiers, or simply because different classifying criteria were used. In any case, we need to know how much error is in our classification in order to use the information we have obtained with some confidence in its accuracy. Ultimately, the information generated by the classification of remotely sensed data will be used to make important decisions about global problems such as deforestation, global warming and environment degradation. It is very important that we not make these decisions based on information that is inaccurate.

An error matrix is the basic tool used for accuracy assessment of remotely sensed data. It gives us a mechanism for generating a number rating the overall accuracy of a classification or map and provides information about the sources of error. This can focus our attention on those areas or classes that require it. We can use this information to improve the quality of our classification criteria and to improve our skill at distinguishing those classes for which there is significant confusion.

Teachers were provided with specific guidance to assist them with advance preparation for the activity. It was suggested that student work sheets be required for each group of two or three students, and that the bird picture sets should be reproduced without the answers on the back. The answers and descriptive identification material would be provided by the teachers as part of the reference data set students would use to determine the accuracy of their bird classification investigation. Materials required for the teacher include the master set of bird pictures, and the master answer (reference data) sheet. We suggested that an overhead of the bird classification worksheet would be helpful in preparing students to use it. Each group of two or three students would be provided the sample beak type sketches, the beak classification work sheet, and the error matrix work sheet.

Following the completion of the activity as presented below, students and teachers are encouraged to discuss the results each group achieved and investigate possible reasons for high or low accuracy among the various reports. Students are encouraged to learn to interpret their data and critically evaluate the results. For example, specific questions suggested to guide this evaluation are: "How did different students' results vary?" "How might we explain these variations?" "What other classifications might be compared using an error matrix?" " How might we use satellite maps identifying land cover for a specific location versus carefully checking the same location in person?"

Teachers and students are encouraged to make the connection to the "big picture" by placing this learning activity in context of the overall science investigation. Following the teachers' presentation of the introductory "global connections" material, they are encouraged to engage students in a "preactivity" discussion to set the stage for the learning activity. Suggested questions include: "Why do we organize or sort objects into groups?" "Name three examples of objects that are commonly sorted into groups." "How do we sort these objects—what are the decision rules we use to categorize things?"

The activity is introduced by providing the teacher with (1) an overview, (2) a glossary of key terms and concepts is included to ensure that participants are clear about the scientifically accepted definitions, (3) a sample error matrix using the bird beak classification system and showing the calculations, (4) a visual representation of the data for use by students who may not wish to use mathematical calculations; and (5) a sample completed Bird Classification Work sheet, with a column showing the validation (correct) data. The teacher provides this validation or reference data to the students after they have completed their independent classification so they can compare their answers with the validation data.

Key Terms and Concepts

accuracy: the degree of conformity to a standard or accepted value. Compare with precision.
The marks on this bull's-eye have high accuracy and low precision:

The marks on this bull's-eye have high accuracy and high precision:

classification: taking a set or group of items and sorting them (classifying them) into well-defined and distinct subsets according to specific criteria. For example, taking a map and outlining areas of evergreen trees, deciduous trees, mixed evergreen and deciduous trees, and nonforest.
criterion: a decision rule. For example, if a forest stand has more than 50% evergreen needles in its canopy, the stand will be classified as evergreen. The preceding definition (c.g., more than 50% evergreen needles) is the *criterion*, the *category* or *class* is evergreen.
data set: a group of values related to the same question being asked. These values will be analyzed together as a group. For example, the set of the heights of all students in this class would be one data set.
error matrix: (see the error matrix on the work sheet at the end of this exercise) a table of numbers organized in rows and columns which compares a classification to validation data. The columns represent the validation data and the rows the classification generated by students. An error matrix is a very effective way to represent accuracy. Correct and incorrect classifications can be compared for each category and used to improve the accuracy of the original classification.
precision: the closeness of several measures to each other. The repeatability of a measurement. This is a very important part of any scientific operation, but is different from accuracy.

> The marks on this bull's-eye have high precision and low accuracy:
>
>
>
> validation data: data collected with a presumed high degree of accuracy. A classification of items (birds in this exercise) is compared to validation data: (1) to improve the decision criteria for the classification (2) to better understand the sources of error in the classification, and (3) to assess the accuracy of the classification data.
>
> Validation data are often collected to improve the classification of an image generated by some form of remote sensing (aerial photography or satellite imagery). Often the term *ground truth* is used instead; however, many scientists prefer the term *reference* or *validation data*. Data that are gathered on the ground always have some degree of error and thus do not represent the "truth."

EXAMPLE EXERCISE

The following is an example of a filled-in classification work sheet, error matrix, and an overall accuracy calculation.

Sample Bird Classification Worksheet			
Bird ID #	Student classification	Validation data	√ or X
1	Carnivore	Carnivore	√
2	Omnivore	Carnivore	X
3	Herbivore	Herbivore	√
4	Carnivore	Carnivore	√
5	Herbivore	Herbivore	√
6	Herbivore	Omnivore	X
7	Omnivore	Omnivore	√
8	Carnivore	Carnivore	√
9	Carnivore	Herbivore	X
10	Omnivore	Carnivore	X

Sample Error Matrix

Validation Data

	Carnivore	Herbivore	Omnivore	Row total
Carnivore	A1. 3	B1. 1	C1. 0	D1. 4
Herbivore	A2. 0	B2. 2	C2. 1	D2. 3
Omnivore	A3. 2	B3. 0	C3. 1	D3. 3
Column total	A4. 5	B4. 3	C4. 2	D4. 10

(left side label: Student Data)

Note: Row and column totals should add up to the same number. Check with others in your group to make sure you counted correctly for each answer in the matrix.

$$D4 = A4 + B4 + C4 = D1 + D2 + D3$$
$$\text{(column total)} \qquad \text{(row total)}$$

How to Read This Information

Across row one (A1–D1) of this example, three carnivores were correctly identified by the students as carnivores, one herbivore was incorrectly classified as a carnivore and no omnivores were classified as carnivores.

Computing the Accuracy

$$\text{Overall accuracy} = \frac{\text{sum of major diagonal} (A1 + B2 + C3)}{\text{total of entire matrix} (D4)}$$

Step 1: Sum the values in the boxes along the major diagonal (A1 + B2 + C3). This number is the total number of correct classifications. In this example there are six correct classifications out of ten total samples.

$$(3 + 2 + 1) = 6$$

Step 2: Divide the total number of correct classifications (A1+B2+C3) by the total number of samples (box D4).

$$6 \div 10 = 0.6$$

Step 3: Multiply by 100 for the overall accuracy of the exercise:

$$0.6 \times 100 = 60\% \text{ accuracy}$$

The calculation can be done for any of the incorrect categories as well (e.g., 3 out of 5 carnivores were classified correctly).

If your answer is between:	*Your level of expertise is:*
0 and 50%	Novice
51 and 85%	Intermediate
86 and 100%	Advanced

The class can also compare fractions (1/2 is less than 3/4, 3/4 is less than 9/10) instead of percentages.

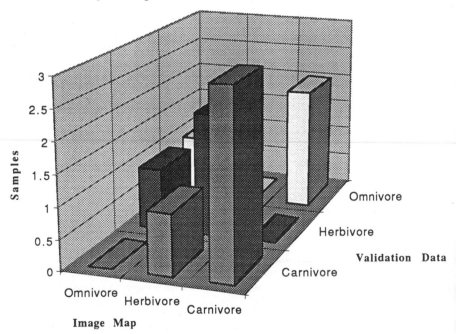

Figure 1. Three-dimensional view of difference/error matrix.

Visual Representation

A visual interpretation can be used instead of mathematically calculating the overall accuracy.

1. Lay out a 3 × 3-cell grid on a sheet of paper numbered like the cells in the error matrix. Visually represent the number of birds in each box by either graphing or physically stacking blocks in the boxes. The tallest columns should be along the diagonal of the grid.
2. If the class has access to computer spreadsheets, a 3-D graph can be created to represent the answers. Figure 1 shows the data from the example error matrix graphed in a 3-D format.

The student exercise is provided below in total, and includes all of the information that the students will need. Also included are questions to help students process their results as well as options for follow-up activities.

STUDENT ACTIVITY: BIRD CLASSIFICATION

Overview

Scientists classify many features in our environment, such as species of life, forest types, or rock types. These classifications, or categories, help us to organize and understand the natural world. For these classifications to be useful to scientists, we need to know how accurate they are. A error matrix is the basic tool used to measure the accuracy of a classification procedure. It also shows us where there was confusion or difficulty classifying certain classes.

In this activity you will:

- Classify pictures of birds into three categories
- Compare answers with the validation/reference data provided
- Generate an error matrix using the results of the comparison

When you have completed this activity, you will be able to:

- Classify birds as carnivores, herbivores, or omnivores using given criteria
- Compare answers to a set of validation data and produce an error matrix
- Identify categories with the most errors
- Evaluate the overall accuracy of the bird classification
- Understand the importance of the error matrix and how to use the information it provides

Materials

1. A set of ten bird pictures
2. Sample beak type sketches
3. Classification and error matrix work sheets

Procedure

In the following activity you will be classifying types of birds as:

C   Carnivores (meat eaters)
H   Herbivores (plant eaters)
O   Omnivores (plant and meat eaters)

Examples of preferred foods:

Carnivores   Fish, meat, insects, worms, small mammals
Herbivores   Vegetation, seeds, nuts, and berries
Omnivores   All of the above

The size and shape of the bird's beak will usually indicate its preferred food type. Many birds are opportunistic, however, and during a food scarcity will supplement their preferred diet with a variety of foods.

Sample Beak Types (Student Reference Sheet for Activity)

Herbivore Beak Types.

Finch type: Heavy wedge shaped beaks are good for cracking nuts and seeds.

Parrot type: Thick curved upper and lower beak are also for cracking nuts or tearing fruit apart. The upper beak has as a sharp point and usually curves over the lower beak.

Carnivore Beak Types.

Insect eater type: Long, slender, slightly curved beaks are used to probe for insects and spiders in tree bark and soils.

Meat eater type: Shorter than the insect eater, upper beak has a sharp curved overhanging tip and straight lower beak specialized for tearing meat.

Omnivore Beak Types.

Jay type: Wide, medium length beak is used for eating insects, fruit, seeds, and even carrion.

Thrush type: Shorter and more slender than the jay type, also for eating meat, plants, and insects.

Bird Classification Work Sheet

Procedure

1. Look at each of the birds on the cards (numbered 1–10) and classify it as a carnivore, herbivore, or omnivore. Record each answer in the student classification column on the bird classification work sheet below.

2. Your teacher will provide the information to be recorded in the validation data column. Be sure to fill in this column accurately; this information will be needed to complete the error matrix.

3. Look at all ten pairs and mark each matching pair with a check mark and each different (incorrect) pair with an "X" in the third column.

Bird ID #	Student classification	Validation data	√or X
1			
2			
3			
4			
5			
6			
7			
8			
9			
10			

4. Fill in the first row of the error matrix by using the following directions:

   A. Count how many times your group matched a student classification of carnivore with a validation answer of carnivore. Place that number here _____. Now place the same number in the box labeled A1 of the error matrix.

   B. Count how many times your group matched a student classification of carnivore with a validation answer of herbivore. Place that number here _____. Now place the same number in the box labeled B1 of the error matrix.

   C. Count how many times your group matched a student classification of carnivore with a validation answer of omnivore. Place that number here _____. Now place the same number in the box labeled C1 of the error matrix.

Be sure to check with your teacher before continuing.

Repeat This Process for Each of the Other Categories Filling in the Remaining Two Rows.

5. Sum the row totals, column totals, and box D4.

Box D4 = A4 + B4 + C4 = D1 + D2 + D3
(column total)   (row total)

The numbers in the outlined boxes (the major diagonal) are classified correctly. Go through the other boxes in the matrix to find any incorrect classifications. The error matrix shows which categories are most difficult to identify.

Which error box has the largest number?

Sample Error Matrix

Validation Data

	Carnivore	Herbivore	Omnivore	Row total
Carnivore	A1.	B1.	C1.	D1.
Herbivore	A2.	B2.	C2.	D2.
Omnivore	A3.	B3.	C3.	D3.
Column total	A4.	B4.	C4.	D4.

Student Data (label on left side)

6. Calculate the overall accuracy as outlined on the sample work sheet.

$$\text{Overall accuracy} = \frac{(A1 + B2 + C3)}{D4} \times 100$$

If your answer is between:	*Your level is:*
0 and 50%	Novice
51 and 85%	Intermediate
86 and 100%	Advanced

Follow-up Discussion and Activities

1. Did you have difficulty correctly classifying a particular category? Why?
2. How could you reduce the number of errors next time?
3. What are some other ways to classify birds?
4. Do you have any suggestions for improving the classification criteria?
5. How did different students' results vary? Compare your error matrix to other students' error matrices to see who had the largest number of accurate answers and to see if other groups made mistakes classifying the same categories. What caused the mistakes?
6. What other measures can be used to evaluate data quality?

Extension Activities

1. Combine all of the class data to create a class error matrix. Calculate the overall accuracy of the class. Which do you think is more accurate, your matrix or the combined class results? Why?
2. Try to develop your own criteria for classifying a group of objects (for example, insects).

Bird Classification Validation Data Sheet

	Bird name	Classification
1	Western greenfinch	Herbivore
2	European starling	Omnivore
3	Bicolored wren	Carnivore
4	Rose-ringed parakeet	Herbivore
5	Brubru shrike	Carnivore
6	Clay-colored robin	Omnivore
7	Pine grosbeak	Herbivore
8	Eurasian jay	Omnivore
9	Common tree creeper	Carnivore
10	Hermit thrush	Omnivore

Bird Pictures (1–6)[1]

[1] All art work on birds by Linda Isaacson.

Bird Descriptions (1–6)[2]

1. Western greenfinch (*Carduelis chloris*)—classification HERBIVORE
This bird (14.5 cm in size) lives in open woodland, bushes, and gardens in Europe, northern Africa, Asia Minor, the Middle East, and central Asia. Its diet consists of nuts and seeds, especially sunflower seeds and peanuts.

2. European Starling (*Sturnus vulgaris*)—classification OMNIVORE
This bird (21 cm in size) lives in open woods, parks, and gardens in Europe and western Asia, and has been introduced to North America, South America, southern Australia, and New Zealand. It eats both plants and animals.

3. Bicolored wren (*Campylorhynchus grieus*)—classification CARNIVORE
This bird (22 cm in size) lives in dry savannah, cactus scrub, and open woods in Colombia, Venezuela, northern Brazil, and Guyana. It finds insects and insect eggs by peering and poking into crevices on the ground.

4. Rose-ringed Parakeet (*Psittacula krameri*)—classification HERBIVORE
This bird (41 cm in size) lives in woodlands and farmlands in central Africa east to Uganda, India, Sri Lanka, and has been introduced to the Middle and Far East, North America, England, The Netherlands, Belgium, and (West) Germany. It eats grain or ripening fruit.

5. Brubru shrike (*Nilaus afer*)—classification CARNIVORE
This bird (15 cm in size) lives in savanna woodland and sometimes the forest edge in tropical Africa. It eats insects and catches food on the wing.

6. Clay-colored robin (*Turdus grayi*)—classification OMNIVORE
This bird (23–24 cm in size) lives in open woodland, woodland edge, and clearings, usually near streams in southeast Mexico, Central America, and coastal Colombia. It eats insects, earthworms, slugs, and lizards as well as fruit.

---

[2] All descriptions adapted from: *The Illustrated Encyclopedia of Birds: The Definitive Reference to Birds of the World*. Consultant-in-Chief Dr. C. Perrins. Englewood Cliffs, NJ: Prentice–Hall, 1990. Art work by Linda Isaacson.

Bird Pictures (7–10)

7

8

9

10

Bird Descriptions (7–10)

7. Pine grosbeak (*Pinicola enuleator*)—classification HERBIVORE
   This bird (20 cm in size) lives in the coniferous and scrub forests of northern and western North America, northern Scandinavia, and Siberia. It eats berries and buds on the ground or in treetops.

8. Eurasian jay (*Garrulus glandarius*)—classification OMNIVORE
   This bird lives in oak woods, and open country in western Europe, across Asia to Japan and southeast Asia. It eats insects, beechnuts, and acorns.

9. Common treecreeper (*Certhia familiaris*)—classification CARNIVORE
   This bird (12.5 cm in size) lives in woodlands particularly coniferous woodlands, in western Europe and Japan. It eats insects and insect eggs gleaned from tree bark.

10. Hermit thrush (*Catharus guttatus*)—classification OMNIVORE
   This bird (15–20 cm in size) lives in woodlands, forest edges, and thickets in North and Central America. It eats insects, spiders, snails, earthworms, and salamanders as well as fruits and seeds.

REFERENCES

Y. Bishop, S. Fienberg, and P. Holland, *Discrete Multivariate Analysis: Theory and Practice* (Cambridge, MA: MIT Press, 1975).

R. Congalton, "A review of assessing the accuracy of classifications of remotely sensed data," *Remote Sensing of Environment.* 37(1991):35–46.

Stephen E. Fienberg, *The Analysis of Cross-Classified Categorical Data* (Cambridge, MA: MIT Press, 1980).

D. Landgrebe and L. Biehl, "An introduction to MultiSpec," Purdue Research Foundation, Purdue University, West Lafayette, IN (1995).

M. Story and R. Congalton, "Accuracy assessment: A user's perspective," *Photogrammetric Engineering and Remote Sensing.* 52(3)(1986):397–399.

CHAPTER 9

Student Data Acquisition

An Important Component of
Student–Scientist Partnerships

Alice B. Krueger, Patrick D. French, and
Thomas G. Carter

From very early childhood, humans acquire information, interpret (or make sense of) it, and make decisions based on that interpretation. However, many adults do not use the most effective data interpretation processes. Nisbett and Ross (1980) investigated the ways adults used anecdotal or database information to make decisions. They found that most people were more influenced by the "vividness" of individual stories than by the statistical soundness of larger samples represented in a database. Kahneman *et al.* (1982) studied a variety of adult decision-making processes and found that many were not based on sound use of data. Even well-trained professional adults, who use acceptable data gathering and interpretation processes in their vocations, do not apply the same processes in their daily lives.

The National Council of Teachers of Mathematics (NCTM) views these disparities between data collection and use in school and in real life as an opportunity to improve classroom instruction. In *Dealing with Data and Chance* (Zawojewski, 1991), NCTM makes two pertinent recommendations. First, "classroom experiences should build on students' natural abilities to use data to

Alice B. Kreuger, Eisenhower High Plains Consortium for Mathematics and Science at the Mid-continent Regional Education Laboratory, 2550 South Parker Road, Suite 500, Aurora, CO 80014; akrueger@mcrel.org.   Patrick D. French and Thomas G. Carter, ADA Technologies, Inc., 304 Inverness Way, South, Suite 365, Englewood, CO 80112; pat.french@adatech.com and gregg.carter@adatech.com.

solve problems in everyday situations of uncertainty." Second, NCTM recommends that "sound reasoning and decision-making experiences with large sets of data should be included in instruction. Students need guidance in learning to interpret and analyze large sets of data, which may seem less vivid or real than smaller sets of anecdotal data."

The *National Science Education Standards* (NRC, 1996) parallel these suggestions in the science content area. The standards recommend "inquiry into authentic questions generated from student experiences" as "the central strategy for teaching science." Teachers should "focus inquiry predominantly on real phenomena, in classrooms, outdoors, or in laboratory settings" as they "guide students in acquiring and interpreting information." In science classes students will "collect data and decide how to represent it... organize data to generate knowledge, and...test the reliability of the knowledge they have generated." They will "assess the efficacy of their efforts—they [will] evaluate the data they have collected, re-examining or collecting more if necessary."

Both the NCTM mathematics standards and the NRC science standards envision gathering, representing, and analyzing data as processes that should be included in mathematics and science courses at all grade levels. The authors of *Making Sense of Data* (Lindquist, 1992) state, "As students are involved in generating questions and collecting data, they will be able to make more sense of data."

Classroom experiences such as the ones available through student–scientist partnerships are valuable in building mathematics and science literacy skills that will be useful to future adults. All of the student–scientist partnerships described in this book involve students in the collection of some type of data. The type of data can vary from numbers of different bird species at a feeder, to cloud types and coverage, to concentrations of particular chemicals in water samples. The timing of data collection can vary from once a year (spotting the first migrating monarch butterfly at a location) to hourly (observing weather conditions). Many types of data can be obtained with unaided human senses (noting colors of pigeons in urban flocks or determining the types of macroinvertebrates present at a field site). Some require simple measurement tools, such as yardsticks to measure snow depth. Other data types require more complex collection technologies: positional plotting devices for following the movements of tagged animals; laboratory equipment such as colorimeters for determining concentrations of molecules present in samples of plant tissues; or portable field equipment for logging temperature, air pressure, and relative humidity information throughout a field trip. Sometimes data can be collected automatically and stored electronically.

USING COMPUTERIZED DATA ACQUISITION SYSTEMS

Collecting and storing data using an electronic data acquisition system (DAS) is commonplace in industry and is finding greater acceptance in schools.

Commercial systems have been designed for every grade level from elementary through high school. Students who use DAS can increase their understanding of the science being studied and also increase their competency with electronic analysis.

Use of these systems has several advantages for both industry and education: regularity, frequency, and accuracy of the data collected. The most important advantage is that a DAS can take data at regular intervals night and day, on weekends, and on holidays. This means that when students and scientists graph the data collected, any trends will be more apparent, because no gaps exist in the data. Generally, a DAS system can take data points at an appropriate frequency from milliseconds to hours apart. In addition, a DAS takes more consistent and accurate readings than a human observer. Lines will be smoother when the data are graphed, which leads to more accurate identification of the systemic causation.

Most DAS can be linked to a computer. The computer allows the user to graph, analyze, and store the data. This can greatly increase the user's ability to visualize the system he or she is trying to understand. Generally, scientific systems being studied are complex and encompass multiple, interacting variables. Statistical correlations among variables can be readily established using computer software packages. To maximize the benefits of the data acquisition, students must learn both the underlying principles and how their experiments prove or disprove hypotheses.

A common misconception regarding use of a DAS is that it is a black box that does all of the work and leaves little for the scientist or student to do. This is totally incorrect. A DAS does take away the tedium of data collection and provide more accurate data, but it certainly does not answer any questions about the system. Instead, what the DAS provides is a "snapshot" of the workings of a complex system. Interpreting that "snapshot" is the responsibility of the scientist or student.

An analogy would be using a computer to draw a graph versus drawing the graph by hand. The computer eliminates the tedium, but it does not answer the original question: What does the graph show? The primary goal is understanding the physical or biological process; learning the mechanics of using the tools is secondary. Similarly, collecting the data is simply one step in the process of understanding the system of interest.

Once data are collected using a DAS, the student or scientist will graph the data so as to recognize any trends. The graph may answer some questions, but usually it generates several new ones. This step in the process is "minds-on" science. Data acquisition is a tool that helps identify the key features of a system; it is not a replacement for human reasoning. In fact, using a DAS can emphasize the need for properly setting up an experiment, performing calibrations, and analyzing the data.

Components of a DAS

Figure 1 shows the elements of a typical DAS. The details may vary, e.g., how each element works and how the DAS is configured. However, some basic features will be found in any DAS.

The beaker in Fig 1 represents a physical system, in this case, a buffered solution. The DAS is measuring the pH, but a sensor could be measuring any phenomenon: touch (soft, hard, smooth, rough, hot, cold, wet, dry); sight (red, green, blue, bright, hazy); smell (cleaning solutions, automobile emissions, food); sound (volume, pitch); taste (sour, sweet, salty); or other property (mag-

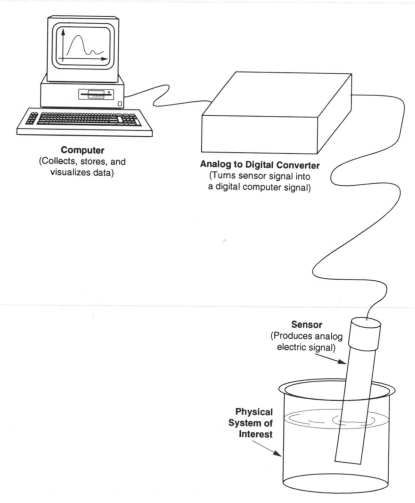

Computer
(Collects, stores, and
visualizes data)

Analog to Digital Converter
(Turns sensor signal into
a digital computer signal)

Sensor
(Produces analog
electric signal)

**Physical
System of
Interest**

Figure 1. Typical classroom data acquisition system.

netic field strength, electric field direction, radiation intensity). The sensor responds to the stimulus by generating an electrical signal. Sensors produce signals that are proportional to their input. This means that a sensor can tell whether or not some quality is present and in what quantity.

In the case above, the pH sensor is generating a voltage proportional to the pH of the solution. Just as the eye can see light but cannot hear sound, a sensor only responds to certain physical inputs. That is, the pH sensor creates a signal in response to pH and not in response to light or sound.

The electrical signal then goes into the analog-to-digital converter. Here, the signal is turned into a digital code which the computer understands. When the sensor's analog signal is converted to a digital signal, it becomes a series of "snapshots" of the original signal. The measurements are no longer continuous, but are sampled at some periodic rate. It is important to collect enough data to avoid missing information that could occur between "snapshots."

The digital signals can be turned into useful information (like a pH reading) by the computer through the use of a calibration equation. This expresses the relationship between the signal the sensor generates and the physical input quantity. Just as our brains convert the qualitative sense of temperature into a roughly quantitative sense of "hotter" or "colder," so the computer translates an input from a sensor into a meaningful number such as the temperature in degrees Celsius. Finally, the data can be stored in the computer, graphed, and analyzed.

Graphing data is integral to many disciplines of science. It is much easier to recognize patterns and understand systems graphically than by reading rows of numbers. Figure 2 shows a computer screen image from a commercial data acquisition package. The data in Fig. 2 were generated by putting a temperature probe in a cup of coffee and collecting data as the coffee cooled. The computer program generated the graph.

By examining Fig. 2, a student would observe that coffee temperature decreases with time. This immediately shows that the room temperature is below the initial coffee temperature. A second observation is that the initial temperature of the coffee is 90°C and the final temperature is 23°C. From this, one can infer that the brewing temperature of the coffee was at least 90°C and the room temperature was no more than 23°C. The next observation would be that the line is not straight, but has a concave curve. This requires a model of why the graph is not a straight line.

Experimenters begin with an initial set of assumptions and a hypothesis. Before we started this experiment, we thought about what we expected to see and why. (Our expectations do not necessarily have to be right.) During this experiment we had expected the coffee to cool down as fresh-brewed coffee is hotter than the room. We initially hypothesized that the coffee would lose heat at a uniform rate, making a straight-line graph. But when we analyzed the results of our experiment, we recognized that the graph is not straight but rather looks somewhat concave. We hypothesize the reason: the rate at which the coffee cools

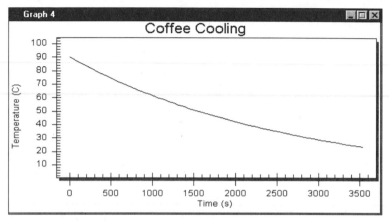

Figure 2. Screen shot from data acquisition software: Coffee cooling.

is proportional to how much hotter the coffee is than the room temperature. This would mean the coffee cools more quickly in the beginning because it is much hotter than the room, and it cools more slowly at the end because it is almost the same temperature as the room. A simple experiment like this leads to more hypotheses and a wealth of learning about scientific reasoning and the nature of things.

EXAMPLES OF DATA ACQUISITION PROJECTS IN EDUCATION

Student Watershed Research Project

The Student Watershed Research Project (SWRP) was developed by Saturday Academy at the Oregon Graduate Institute (OGI), Portland, in 1991 with NSF funding. It involves 8th to 12th grade students in collecting water quality data from local streams. Teachers implement the curriculum in two sections: 6 weeks in the fall and 4 weeks in the spring. Near the participating school, two field sites are identified, often one upstream and one downstream on the same body of water. A 5-day summer training prepares classroom teachers to work with the project's six modules. The project requires two to four field trips during the year to sampling sites and a fifth to the Annual Student Watershed Summit, where students present their findings and interact with students from other participating schools.

Teachers choose from six modules: Basic Chemistry, Advanced Chemistry, Microbiology, Macroinvertebrates, Vegetation Inventory, and Habitat Assessment. These modules require different levels of observational technology, from simple identification to the use of colorimeters and microbiological techniques. For Basic Chemistry, students collect water samples at the two different field sites after recording the temperatures of the samples at the sites. The students use the protocols of Hach kits to test their collected water samples for pH, alkalinity, and biological oxygen demand (BOD). If the school has the equipment, students also can measure water turbidity and total solids.

Two modules require more sophisticated collection and analysis equipment. In the Advanced Chemistry module, students use a commercial colorimeter to measure the amounts of phosphate, ammonium, nitrate, and chloride in their water samples. This process requires training in the analysis of synthetic unknowns prior to sampling in the field. Laboratory testing helps improve student precision, troubleshoot potential errors, and teach quality control procedures. Two of the Advanced Chemistry tests involve standardized waste collection and disposal techniques. In Microbiology, water samples are plated and examined for the presence of fecal coliform bacteria. Students learn to sterilize equipment and prepare agar plates prior to sample collection.

Three modules require visual observation of living and/or nonliving ecosystem features. In Macroinvertebrates, visible invertebrate animals such as insects and planaria are documented. The insects are sorted, counted, and classified to taxonomic order. An indication of the health of the stream can be extrapolated because previous studies have identified the percentages of insect orders found in healthy environments.

In the Vegetation Inventory module, students record the presence of various plant species. Students collect samples of vegetation to press and mount on herbarium paper. The herbarium record is periodically submitted to the biology department at Pacific University for verification of plant identification. In Habitat Assessment, students use 11 parameters to analyze the suitability of a stream for fish. Students work directly in the stream as they determine substrate composition, stream flow, stream depth and width, channel shape, lower bank stability, bank vegetation protection, in-stream cover, and other factors. The assessment is based on procedures established by the U.S. Environmental Protection Agency for monitoring streams in the Northwest. Students enter their collected data into a customized data entry program, then generate reports. Participating state and local agencies use student-collected data to develop regulatory criteria and to determine which streams need cleanup work.

SWRP's data quality control plan begins with teacher training during summer workshops. Teachers receive equipment and technical support from the project staff throughout the year. Students working on the Advanced Chemistry module test unknown synthetic samples in the lab before analyzing samples from the stream. This improves students' skill and accuracy with the equipment and procedures. At the site, students collect duplicate samples, one set to process themselves and one to be analyzed by a participating quality control laboratory. The results from the lab are listed next to the student-collected data. All student-collected data are audited by project staff to sort out any inconsistencies prior to developing reports for participating agencies.

During the 1996–97 school year, SWRP will work with 22 teachers from 21 schools in the Portland–Vancouver metropolitan area. The project will involve approximately 1000 students, collecting data at 43 sites on 21 streams. This data collection and analysis project has engendered great enthusiasm among student and teacher participants. Lenore Simpson, a Sunset High School (Portland) science teacher, says, "SWRP has provided us with a marvelous integrated unit. The unit has students collecting data on plants, invertebrates, chemistry, and habitat assessment. In a time when educators are encouraged to make learning integrated, relevant, and 'real world' focused, I can't imagine a program which has accomplished these goals more successfully."

Particles and Prairies: The Fermilab Prairie Savannah

The site of Fermi National Accelerator Laboratory (Fermilab) near Batavia, Illinois, was originally a prairie. The initial restoration project in 1974 involved 9.6 acres, 100 volunteers planting native flora, and an institutional vision. Plantings by volunteers continued over the years, with occasional controlled burnings which further establish the prairie species. In 1984 Argonne National Laboratory ecologists started to conduct research on the restored prairie.

When more than 700 acres of prairie had been restored in 1988, Fermilab was established as a Department of Energy National Environmental Research Park. At that time it was the largest reconstructed prairie in the world. Today it is unique because of its many samples of native prairie life forms including a herd of bison. It also contains an Interpretive Trail and the Lederman Science Center.

An educational program, Particles and Prairies: The Fermilab Prairie Savannah, was established in 1991. In this collaborative project, educators and scientists provide middle school students with opportunities to do prairie ecology science. While gathering data and making observations at the restored prairie site, students are not only doing hands-on science but also contributing vital data toward the understanding of a native ecosystem.

Teachers prepare for their class visit to the prairie site by attending a 4-day summer workshop that explores the many educational opportunities available outdoors and inside the Lederman Science Center. The participating teacher will purchase a teacher's guide and can obtain a classroom kit containing measurement tools, field guides, and related multimedia materials (video and slides). The teacher then designs a field and laboratory experience that will benefit students during their visit to Fermilab.

During the outdoor field session, the students collect data on various biotic and/or abiotic features of the prairie ecosystem. The data collected may be in the form of quadrant population studies; species counts; herbarium collections; measurements of temperature and chemical characteristics of soils; or measurements of temperature, humidity, wind speed, and light level above the ground. Docents, who are retired teachers or scientists, are trained to lead student data collection tours.

When the students return to the computer lab, they enter their data into a specially designed spreadsheet program. Their data are combined with historical data from both scientists and other visiting classes. Prairie ecologists utilize these data to further their understanding of the restoration process. During the 4 years of Particles and Prairies operation, approximately 13,000 students from grades 6 to 8 have contributed data to this effort.

An interactive video called Particles and Prairies is available in the Lederman Science Center. It contains 37 video segments and 10 slide collections, which show the prairie throughout its history, in different seasons, and as a research site. The video won the Award of Excellence and the Gold Award in the Education, K–12 category in the NewMedia INVISION 1993 Multimedia Award Competition. Software to explore the information in this video is available at a multimedia station in the Lederman Science Center.

Additional Examples

Winter Weather and Whitetail Deer involves students in collecting snow depth data with a yardstick. This information is recorded daily on a data collection sheet which is mailed or faxed monthly to the New Hampshire Fish and Game Department, the project's sponsor. There it is compiled with data from school sites across the state. Fish and Game statisticians calculate a county-by-county Winter Severity Index. These indices are the basis for wildlife management decisions made by Fish and Game biologists.

Deer and other foraging mammals are affected by both snow depth and air temperature. In colder weather, deer need to obtain more food to maintain their body temperature. With deeper snow, food is harder to find, and more deer will die. The number of deer likely to be able to survive during a certain month can be calculated, and hunting can be increased or decreased through the sale of licenses so as to maintain the proper size of the deer herds.

Students receive a monthly summary of all snow depth data submitted and an annual summary report showing the monthly derived Winter Severity Indices and deer harvest reports. The relationship between the data submitted and the number of animals harvested by hunters can be explored. A curriculum guide is available for teacher use. In the 3 years that this program has been functioning, approximately 30 schools have consistently contributed data.

Project SUN (Students Understanding Nature) is sponsored by the Jet Propulsion Laboratory (JPL) in Pasadena, California, and by the National Aeronautical and Space Agency (NASA). High school students use specially modified computers to collect intensity data on visible and ultraviolet sunlight, correlated with local weather conditions such as temperature, cloud cover, and humidity. Researchers from National Renewable Energy Laboratory (NREL) and NSF have validated the sensor designs. A team of volunteer students collects data using automated data logging devices every 3 minutes from 10:00 A.M. to 2:00 P.M. Single purpose software compiles the collected data, requests input of local weather conditions, and creates a final text file, which can be submitted to JPL over the Internet or on disk through the mail.

Teachers initiate this educational program by bringing their earth science classes to an assembly conducted by the project's director. He shows the types of instruments used in his research and discusses some research results. When a team of students expresses interest, the director gives them a 2- to 3-hour hands-on session with the project equipment. The teams of students are required to submit data to the project 2 days every week, and they may use the equipment for their own research the rest of the time. Many students produce science fair projects related to product safety, such as testing sunblock lotions or sunglasses.

This project has been in operation for 1½ years and has ten participating schools, some in California and others in Australia. In addition to NASA scientists who access the database, German researchers are using the SUN data in their efforts to discover soil additives that can increase a plant's resistance to the effects of ultraviolet radiation. The SUN data collection system is in place in Hawaii at the summit, on the wet side, and on the dry side of the Mauna Kea volcano, to study microclimates for a future project.

INDUSTRY'S TOOLS OF TODAY MEAN EDUCATIONAL USE TOMORROW

There are two major reasons why current industrial techniques of data acquisition and control will become educational uses in the future. First, use of data acquisition and control achieves significant cost savings in industrial applications. This means that resources can be devoted to developing and purchasing new tools for further cost savings and quality improvement. When these tools become commonplace in industry, their prices drop significantly, which then makes the tools cost effective for education. Second, as various data acquisition tools become common in industry, the future work force will be expected to learn how to utilize them.

DAS Helps Utility Industry Control Pollution

The applications for data acquisition and control vary greatly from industry to industry. A coal-fired power plant is the subject of the following example. The goal of a coal-fired utility is to generate low-cost electric power safely with minimum pollution. This is not a particularly easy task; it requires a constant balancing act between operating parameters.

Burning coal releases heat energy and effluent gases. Quantities of pollutants depend on conditions such as type of coal, temperature of the furnace, available air, and so forth. Sometimes additives are used to control potential pollutants. The additives cost money and also may generate other pollutants if used in the wrong quantities. Finding the right operating parameters might be easier if the demand for electric power were constant. However, electricity demand changes greatly over any 24-hour period, so power plants are constantly adjusting power generation to users' needs. Every change in power generation requires adjustment of all of the pollution control systems. Obviously, this process would not be possible without automation features.

In the control room of a power plant is a myriad of instruments, which monitor and direct the power generation and pollution control processes. Each one of these instruments collects information from sensors, stores that information, allows data visualization, and provides alarms. With this information, on-site operators make adjustments to each process. The ultimate control rests in the hands of the operators.

ADA Technologies, Inc., a Colorado company, conducts a significant amount of pollution control research. The company often conducts experiments by placing pilot, small-scale pollution control devices at power plants. One example is a structure about the size of a semitrailer truck called a baghouse. It contains dozens of sensors monitoring gas flow, temperature, pressure, and other variables. These sensors feed back to one central computer, which is equipped with a custom software program that monitors and controls the baghouse. For example, the computer controls overall effluent gas flow by taking the information from the flow sensor and using that information to adjust a flow control valve up or down (Fig. 3). Similarly, the temperature of the unit is controlled by the computer, by examining data from a set of temperature sensors and then deciding which heaters to turn on and off. This same pattern is repeated for dozens of variables, creating a controlled experiment. The overall system can appear very complicated, but actually is a set of very simple components that read a variable and determine an appropriate response. The appropriate responses have been programmed into the software prior to an experiment involving several variables.

In addition to monitoring the entire system and making periodic adjustments, the computer also stores information and transmits it by telephone to ADA Technologies's central office. This information is reviewed daily in graphic

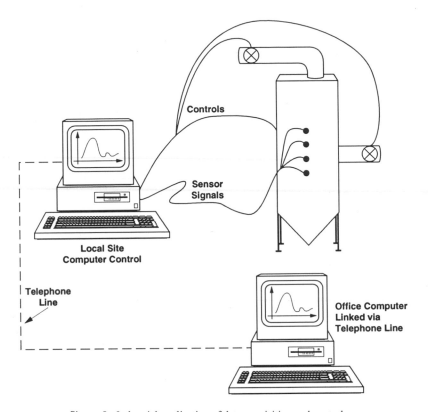

Figure 3. Industrial application of data acquisition and control.

formats, allowing easier interpretation of the system performance. Based on the daily results, decisions are made about procedures for the following day. On occasion various elements in the system break down, or unexpected circumstances occur. These problems are generally diagnosed using the same sensors. In these cases, an off-site operator will adjust a valve (or other component) via the phone lines and watch the real-time results. A visit to the site may or may not be required.

In general, industrial applications of data acquisition include

- Monitoring processes using sensors
- Controlling processes through actuators such as valves
- Programming systems using computers
- Communicating with equipment via telephone
- Storing information for future use
- Providing alarms to users when difficulties are encountered
- Performing internal calibration checks

Today's engineers use tools that include sensors for hundreds of physical parameters, data collection tools, easily programmed software, communication tools, and computers. In the future these tools will become as commonplace in schools as they are in industry, although there will always be a time lag between equipment used in industry and that used in education.

Classroom DAS Use Today

In middle and high schools today, the largest use of DAS equipment is in physics classes. More products are available to this market because physics teachers tend to be more interested in fairly sophisticated electronic equipment. However, a relatively small portion of physics classes actually have data acquisition equipment to use.

Currently, little easy-to-use equipment is found in chemistry, biology, and environmental science classes and only small companies are manufacturing this equipment. As teachers become more knowledgeable about DAS uses, they may see reasons to obtain equipment for use in nonphysics courses. As more classrooms begin to use these tools to stimulate problem solving skills, the market will grow. At that point, more companies will start selling lower-cost, user-friendly instruments to the schools.

Factors that have limited the use of these tools in classrooms include lack of computers in schools, lack of curricular units that use the equipment, lack of teacher training, and, to a lesser extent, the cost of a DAS. The number of computers in schools has increased constantly over the last decade. In 1983 there was about one computer for every 125 students in U.S. public schools. By 1995 there was a computer for every 9 students. This trend will continue and will eventually cancel one of the major factors limiting data acquisition in classroom activities. Curricular units and teacher training will remain limiting factors. However, as computers become more commonplace, the general level of teachers' computer literacy will increase. Cost will always be an issue, and companies that want their equipment to be cost-effective for schools will need to make it usable in several disciplines.

One example of teacher and classroom acceptance of a DAS is a product being developed by ADA Technologies called LabPen. This is a portable data collection device with "plug-and-play" sensor modules. During development of the product, the company conducted several field tests to see if the tool was appropriate for different science disciplines. Once teachers were shown the instrument's capabilities and simplicity of operation, they quickly embraced the technology. At one field test location the equipment was tested in several classes beyond those in the initial plan. When teachers showed each other the capabilities of the equipment, they began using it in all science disciplines. At another school, testing was scheduled for two physics classes only. However, when the equipment was returned, students in several chemistry classes had filled out evaluation forms as well. Once new equipment finds a way into a school, collaboration among teachers leads to other uses for that equipment.

Field testing of LabPen also was successful in a class that included special education students. This particular environmental science class included some low ability students meeting a graduation requirement plus some mainstreamed special education students (learning disabled, mildly mentally challenged, and behavior disordered). The class was cotaught by a certified science teacher and a special education teacher. The science teacher was pleased that all of the students were able to obtain valid field data and create graphs. The special education teacher appreciated the ease of students' manipulation of the data collection equipment and the comprehension they showed during data interpretation.

Another method for data acquisition in the classroom uses the Internet. Schools are rapidly coming "online" and can communicate easily with other schools and scientists. Several groups, quick to recognize this potential, have started networked learning experiences, Some of these have students around the world collect data and transmit them via the Internet. These student–scientist partnerships include the GLOBE Program and Kids as Global Scientists. Although most of these networks collect data by hand at present, electronic data acquisition could be the next step as communications, organization, and equipment improve. Some projects, such as Athena, make use of preexisting databases that are available through the Internet.

Hardware Choices Currently Available for Education

Some of the products available for data acquisition in science and mathematics courses are the following:

- *Calculator-Based Laboratory* (CBL) is a hand-held, portable, battery-powered data logger. These data can then be downloaded to a Texas Instruments graphing calculator. A series of compatible sensors is available through Vernier Software. The CBL is produced by Texas Instruments. For more information contact a Texas Instruments distributor.

- *LabPen System* is a hand-held, portable, battery-powered data logger with four ports for self-identifying sensor modules. LabSpread, the associated software, downloads data, has an analysis function, and generates graphs. The LabPen system is not yet commercially available. Contact Judith Armstrong of ADA Technologies, Inc., Englewood, Colorado, for more information, 303-792-5615.

- *Personal Science Laboratory* (PSL) is an interface for PC- or Mac-compatible computers. PSL was originally produced by IBM. Its highly accurate probes are warranted for 3 years. An Excel-based software package and an ward-winning curriculum including an Excel experiment workbook are available. For more information call Team Labs in Boulder, Colorado, 800-775-4357.

- *Quantum Leap* is an interface box (an MBL) between sensors and either a PC- or Macintosh-compatible computer. Software and sensors are available through Quantum Learning Technology of Boulder, Colorado, 303-674-9651. The company also produces a hand-held data logger called the *Trekker*.
- *Science Workshop 300* is a portable interface with software for Mac or Windows, which takes data through three analog channels. It plugs into the serial port of Windows-compatible computers or into the modem port of a Mac. Nineteen types of compatible sensors for use in biology or chemistry are available from PASCO of Roseville, California, 916-786-3800.
- *Sense and Control* is a data logger that can operate in conjunction with or independent from a computer. The multichannel unit has no display. Software for either Macintosh- or PC-compatible computers and a series of self-identifying sensors are available through Data Harvest of Grand Island, New York, 800-436-3062.
- *SensorNet* is a sensor interface box that connects to a Macintosh computer and includes software. Sensors are available for general science, environmental science, chemistry, biology, physiology, and physics classes. SensorNet is available from the AccuLab Products Group of Modesto, California, 209-522-8874.
- *Universal Lab Interface* (ULI) is a series of products for use in laboratory settings, which communicates with either a Macintosh or PC computer and a host of sensors. Software and sensors are available. The ULI is produced by Vernier Software of Portland, Oregon, 503-297-5317.

VALIDATION OF STUDENT-COLLECTED DATA

For a student–scientist partnership to function properly, two factors need to be present. Students and their teachers need assurance that the educational aspects of the data collection will not be sacrificed to the need for validated data. Scientists need to know that the students are collecting accurate data that have been scientifically validated.

A validation method checks either the precision of the measurement technique or the accuracy of a specific collected datum. One method to improve the accuracy of students' data-gathering techniques is to assign only one measurement to an individual student. A single student is trained to make the one measurement very precisely, which can provide reasonably accurate data. This approach is generally the one scientists prefer, but the entire class learns little of the overall aspects of making measurements.

The previous method tends to have educational shortcomings, but another approach puts the educational value first. Every student has at least one opportunity to make each type of measurement. This leads to a better educational experience, but the data may be of significantly lower accuracy and not validated. The experience could be diminished for the students if the data are not gathered for a scientific research project but are merely another classroom exercise.

A third approach is to validate data while retaining educational value. Two research projects, one Canadian and one in the United States, have addressed this approach in different manners.

Canadian Project Addresses Data Validity

The Niagara Escarpment Biosphere Reserve Project–Community Outreach through Environmental Monitoring (NEBR'S Network) is a Canadian project addressing validity of student-acquired data with a direct research project. Permanent outdoor sites which have been established for 25 years and which will be available for research into the foreseeable future were selected for this research. Two field plots are set up in similar sites in the same ecosystem, a maple-beech closed canopy forest. The control plots are accessible to scientists, on a permit basis only. The demonstration plots are available to students, local landowners, and scientists. Logs are kept to track use and for subsequent impact studies.

The concept is to use both low- and high-tech data collection equipment, with data collected by members of the community at large and by the scientific community, to monitor changes in the environment. The purpose of this research is to compare the data taken by the two groups. Identical research is conducted on the two plots. Data collection on tree growth uses Smithsonian protocols. Data collection on ultraviolet B, amphibians, earthworms, phenology, and climate will be undertaken as protocols are established. On one plot all measurements are made by scientists; on the other, identical techniques are used by students ages 10 and up. Participating schools are recruited from those which visit the local Outdoor Center for a field experience. Students and their teachers are trained in data collection protocols on the site at the Outdoor Center. Scientists are invited to collect data at the site by the project's Technical Advisory Committee, which is responsible for the scientific assessment of the overall project.

This project is funded by the Federal Environment Canada Action 21 program, as well as in-kind donations from 12 partners. The partners include two school boards, the Royal Botanical Gardens, the Niagara Escarpment Commission, and four Canadian businesses whose goods and services are vital to the project. Other partners are the Council of Outdoor Educators of Ontario and ACER (Association for Canadian Educational Resources). NEBR'S Network was initiated in July 1996. It will conclude in July 1998, when a self-sustaining NEBR'S Network is in place. Information is available from the project director,

Alice Casselman, president of ACER. (Please e-mail requests for information to acercass@astral.magic.ca)

Software Could Ensure Both Data Validity and Good Pedagogy

A project in development aims to validate student-collected data without sacrificing educational aspects of a student–scientist partnership. This project is developing hardware, software, and classroom pedagogy. The hardware is a data logger along with a set of self-identifying sensor modules. Each data point collected will include the reading, the serial number for the specific sensor probe used, the calibration constants used, the time and date the data were collected, and the identification number of the student who collected the data.

Collected data pass to software that checks for consistency with related historical data. Data that are not validated are returned to the originating school for further analysis, consisting of two tests. First, the calibration of the particular sensor is manually checked. Second, the individual student's methodology for that type of measurement is reviewed. Thus, instruction in technique can meet individual student needs. Inspection of a data set for validity is an important lesson for the entire class.

The flow chart in Fig. 4 illustrates the procedure to be used for validating student-collected data. Starting at point A, when data are received by the network software, they are checked against the physical limitations of the hardware. Often, when a probe is broken, it reads either at or above the upper limit or at or below the lower limit. Next, B, a "statistical best guess" as to what the reading should have been, given historical data, time, location, and nearby locations, is calculated. The "statistical best guess" becomes more and more accurate as the project continues to gather more data. The predicted value is calculated with a standard deviation, indicating how precise the range is expected to be. This standard deviation should get smaller as more data are collected.

Once the predicted value is calculated, it can be compared to the current reading, C. If the current reading falls outside a preset range around the predicted value, that data point will be flagged. Next, the specific probe history is checked, D, to see if this probe always reads above or below the predicted value, but not enough to be outside the accepted range. This test can recognize a probe that is consistently low or high, no matter at what location the probe is used. A confidence number is generated for that specific probe. It is important to remember that if a data point is out of the accepted range, it is immediately flagged for evaluation.

The next test, E, is very similar to D, except that it examines each student's historical performance in making a specific measurement. Students will be using different probes every day and will be making measurements at different locations, but it is still possible to test their ability to make an accurate measurement. If a student's readings are consistently higher or lower than the predicted value

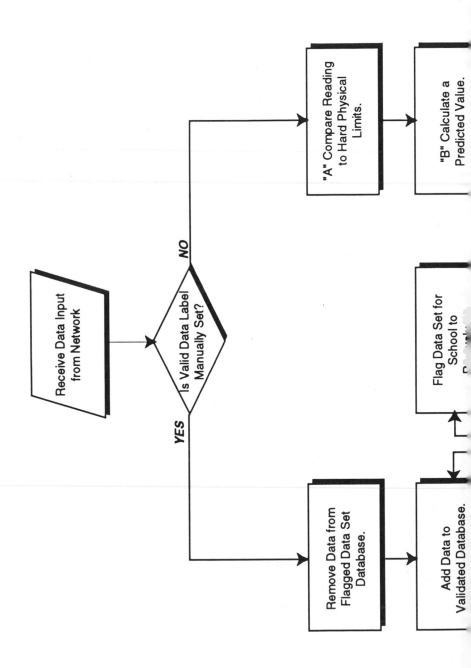

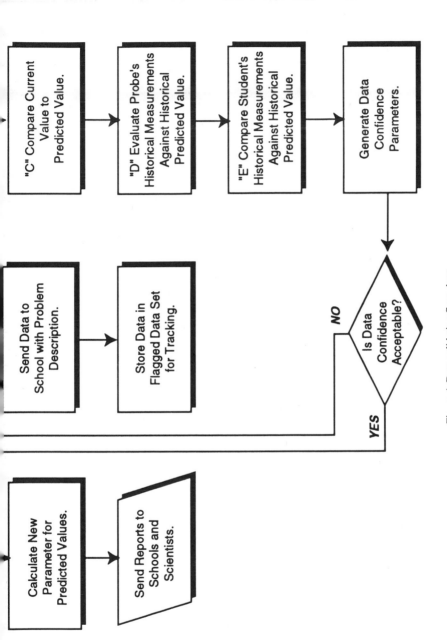

Figure 4. Data validation flow chart.

(but within an acceptable range), then a subtle problem may exist. A confidence number for the student's ability to make that specific measurement is calculated. If a data point is found to be within the appropriate bounds, it is accepted into the validated database. If it has not been accepted, it is stored in the flagged database and returned to the school that submitted it. Flagged data undergo two further checks, after which they are either discarded or returned to the accepted database. The first test is of the hardware; the probe's calibration is checked. The second test reviews the student's technique for that particular measurement. Because the validity of each data point can be checked, the project can have many students performing several types of measurements, providing a more varied educational environment and allowing the students to understand the larger picture of the project. This method also ensures the validity of the data gathered, providing both accurate data to scientists and a meaningful learning experience to students. For information about this project, contact Patrick French at 303-792-5615.

Using computers to identify questionable data has three distinct advantages. First, all of the data are eventually individually validated, and the database consists of very consistent, accurate data. Second, most of the data can be immediately validated so that effort is concentrated on the more important scientific issues. The third advantage is that only the probes and the students who provide statistically questionable data need to be examined. This offers a feedback that can be used to improve students' data collection skills and to question outlying data points. Flagged data should not be automatically discarded. At times, the outlying points actually clarify some scientifically important factors.

Most of the student–scientist partnerships in existence are not well developed enough to make immediate use of such a data validation tool. By the time these new tools have been fully developed and field tested, however, many student–scientist partnerships will be ready to use them.

REFERENCES

D. Kahneman, P. Slovic, and A. Tversky (eds.). *Judgment under Uncertainty: Heuristics and Biases.* (Cambridge: Cambridge University Press, 1982).

Mary M. Lindquist, with J. Luquire, A. Gardner, S. Shekaramiz, and consultants D. Clements, J. Firkins, J. Joyner, *Making Sense of Data.* (Reston, VA: National Council of Teachers of Mathematics, 1992).

National Research Council (NRC), *National Science Education Standards.* (Washington, DC: National Academy Press, 1996).

R. Nisbett and L. Ross, *Human Inference: Strategies and Shortcomings of Social Judgment.* (Englewood Cliffs, NJ: Prentice-Hall, 1980).

Judith S. Zawojewski, G. with Brooks, L. Dinkelkamp, E. D. Goldberg, H. Goldberg, A. Hyde, T. Jackson, M. Landau, H. Martin, J. Nowakowski, S. Paull, A. P. Shulte, P. Wagreich, B. Wilmot, and consultants D. Kroll, Lambdin, F. Lester, Jr., K. Goldberg, *Dealing with Data and Chance.* (Reston, VA: National Council of Teachers of Mathematics, 1991).

CHAPTER 10

Role of Technology in Student–Scientist Partnerships[1]
Perspective from the National Science Foundation

M. Patricia Morse and Nora H. Sabelli

INTRODUCTION

A variety of student–scientist partnerships (SSPs) have evolved over the years in the K–12 classroom, providing students, teachers, and scientists with activities and experiences not available from any other source. Characteristically these activities are inquiry-based or design-oriented and aligned with *Benchmarks* (American Association for the Advancement of Science, 1993), *National Science Education Standards* (National Research Council of the National Academy of Sciences, 1996a), and numerous state frameworks for science education that call for students to understand the process of inquiry.

Although inquiry-based activities without technology are possible, the scope and nature of these activities are heightened by the use of computers,

---

[1] Any opinions, findings, conclusions, or recommendations expressed in this chapter are those of the authors, and do not necessarily reflect the views of the National Science Foundation.

M. Patricia Morse, Northeastern University, Marine Science Center, Nahant, MA 01908; *present address:* Division of Elementary, Secondary, and Informal Education, National Science Foundation, 4201 Wilson Boulevard, Arlington, VA 22230; mpmorse@nsf.gov.   Nora H. Sabelli, Division of Research, Evaluation, and Communication, National Science Foundation, 4201 Wilson Boulevard, Arlington, VA 22230; nsabelli@nsf.gov.

research technologies, and the Internet (Jackson *et al.*, 1994; Sabelli, 1994; Jonassen and Reeves, 1995). As reflected in recent publications, different technologies have become increasingly important in driving new questions, new activities, and new forms of collaboration in the K–12 environments (for general information, see Weir, 1992; Horwitz and Barowy, 1994; Julyan and Wiske, 1994; Feldman and McWilliams, 1995; Means and Olson, 1995; Office of Technology Assessment [U.S. Congress], 1995a,b; Berenfeld, 1996; Fulton *et al.*, 1996).

The National Science Foundation (NSF) supports research in and use of educational and research technologies in K–12 classrooms. Ongoing funded projects in the Directorate for Education and Human Resources, especially in the Divisions of Research, Evaluation, and Communication (REC) and Elementary, Secondary, and Informal Education (ESIE), address the use of educational technologies in research about children's learning, in prototypes for technology-based materials for science, mathematics, and technology education, in development of instructional materials for inquiry-based curriculum, in networking and use of technology for learning science and mathematics in the informal settings of museums and science centers, and through the application of new ways to enhance teachers' knowledge and skills.

A current interest of NSF Directorates, called for in the NSF strategic plan (1995), is the integration of research and education. Related to this strategic plan is a new initiative developed across the NSF Directorates to fund research and development on Learning and Intelligent Systems, which is an effort to build a stronger scientific understanding of the base for learning, including how to use technologies and technology-rich environment to support learning of complex concepts and ideas (see NSF 97–18 and http://www.nsf.gov/lis). Many innovations spawned by existing programs have found and will continue to find direct application to the activities associated with SSPs in the K–12 environment.

When students, scientists, and teachers truly collaborate, they build a powerful environment for learning science, whether in the K–12 classroom or in an informal setting. The Internet has been particularly useful in facilitating collaborations that bring global perspectives and scientific depth to students' work. It is important to bring scientists, teachers, and students together, based on the complexity of the information that is increasingly becoming available and needed by individuals in our society and the increasing desire for education to enable wider transfer and use of that information in productive ways.

In this chapter, we will discuss the role of technology—and of NSF-supported research and development in the use of advanced applications of technologies—in the creation of research models for learning in the context of partnerships. We focus on (1) inquiry and design, and how these align with the national standards, (2) evolving roles of research and communication technologies education in SSPs, and (3) initiatives and programs at the NSF that were fundamental in establishing the research and development base for many of the activities, and that respond to their need and foster their expansion. We address such questions as why *all* students must engage in inquiry, why students need to have access to current science knowledge, why scientists are vital participants in classroom-based inquiry, and what the implications are for teachers and researchers of partnerships in the classroom.

To succeed in creating SSPs, we need to listen to both the K–12 and the scientific communities, and encourage innovative pathways and mechanisms for funding such partnerships. Activities may take the form of increased integration of discoveries from science, mathematics, engineering, and technology disciplines into curriculum development, new ways for preparing teachers and scientists to participate in productive partnerships, research on student learning outcomes, and research on innovative ways to infuse technologies into classrooms in support of the activities mentioned. The SSPs provide a challenging paradigm for education.

INQUIRY AND DESIGN; THE INFOSPHERE AND COMPLEXITY

When discussing how scientists, engineers, and mathematicians might interact with students, we are reminded that both *inquiry-based* and *design* activities can take place. As discussed in the National Science Education Standards (NRC, 1996a), these activities complement each other and are related to the *natural* world and to the *built* world. Collins (1996) notes, "Inquiry is more than the familiar processes of science. It includes these processes used with deep knowledge of science subject matter, reasoning ability, and utility in the application of science understanding. Inquiry implies facility in asking questions about nature, conducting investigations using appropriate tools and techniques, constructing and analyzing explanations, and communicating arguments" (p. 35). As stated in the National Standards, "Students at all grade levels and in every domain of science should have the opportunity to use scientific inquiry and develop the ability to think and act in ways associated with inquiry, including asking questions, planning and conducting investigations, using appropriate tools and techniques to gather data, thinking critically and logically about relationships between evidence and explanations, constructing and analyzing alternative explanations and communicating scientific arguments" (p. 105). Complementary to this, students working in a scientific enterprise should have the opportunity to identify and state problems, design and implement solutions including estimates of cost and risk–benefit analysis, evaluate their solutions, and present their conclusions to others (Raizen *et al.*, 1995). Thus science as inquiry—as a way to know and communicate about the natural world and solving problems taking into consideration constraints—as a way to benefit people in the built world—is encouraged by activities associated with SSPs.

Before we consider the impact of inquiry-based activities and the increased use of technologies by students, we should examine the reasons that justify them as more than extracurricular activities for a few interested individuals. We need to consider that public education and mass literacy are 20th-century phenomena that are closely tied, as Marshall (1993) has so persuasively argued, to the manufacturing industry needs earlier in the 20th century. As we enter the 21st century, the equivalent mass literacy can be surmised to include technological literacy, the ability to pose and solve problems, and the derived widespread benefits from easily available information.

Students are entering a world characterized by an increased use of computers, automated instruments, and communications technologies in both the workplace and the home. These computer and communications technologies have enabled the expansion of our capacity to solve problems that were thought of as intractable only a short time ago, and are coupled with the concomitant integration of technologies into the industrial workplace. The need for related knowledge and technological fluency among both employers and employees has expanded enormously, and will continue to do so for the foreseeable future.

In addition, given the increased open access to large amounts of information (not just in science), the ability to validate that information and use it appropriately is a new skill that requires nurturing. In the past, this information was only available to professionals in specialized libraries, or was preselected and summarized by intermediaries such as authors, editors, and publishers. With the Internet and increased access to libraries and other on-line resources, the information is available to all students (for example, see Follansbee, 1996). Processing and validating information and constructing their own knowledge on the basis of information becomes important for students, and the scientist partners can contribute and help the teachers and students in this process. Berenfeld (1996) describes this global environment as the "infosphere," likening its importance to the Earth's biosphere, atmosphere and lithosphere. He furthermore characterizes the infosphere as a cohesive and integrated flow of knowledge that reflects a current political and cultural changes in a timely manner, and encompasses all communications, media (including the newer computer-mediated communications), computing simulations, and virtual reality (Berenfeld, 1996).

For students to build the capacity to address interesting and challenging problems, among other things, they must develop the ability to deal with issues of *complexity*, such as nonlinearity, multilevel system organization, feedback, and chaos. These concepts are fundamental to the understanding of increasingly relevant—socially and scientifically—concepts and phenomena. Air pollution, solar energy, the AIDS virus, water purity, acid rain, genetic manipulation, biodiversity, and many others are concepts of modern daily life, and their "complexity" makes these concepts seem daunting. This is particularly difficult for teachers in the K–12 environment, when they contemplate working with *existing* educational tools and their own expertise. A classroom-only focus obscures the fact that we *can* make complex ideas "supportable, manipulable, and understandable" by collaboration with scientific experts and using appropriate tools.

The gap in communications between scientists and the public, independent of support and tools available to teachers in the classroom, still exists. Narrowing this gap might enable students and the surrounding public to understand the scientific concepts behind complex systems and should guide long-range curriculum policies. Albert Einstein warned that everything should be made as simple as possible, but not simpler. For students, as future citizens, to act responsibly on science's significant discoveries, they must learn to simplify and extract meaning *without* oversimplifying complex issues to where the issues become less insightful and powerful.

Evolution of research and education technologies occurs at a rapid rate. As the requirements of research in science, mathematics, and engineering have evolved, more powerful computers and higher bandwidth networks have been developed to serve them. These technologies have made many scientific and engineering advances possible. Moreover they have dramatically changed the public access to shared information and changed the scope of interinstitutional collaborations. Users are creating their own networks, formulating their own goals and procedures, and creating, manipulating, and sharing information and data with others with like interests and without concern for distance. This provides an opportunity for new and different forms of remote access to data and equipment, and for collegial and expert collaboration that could substantially improve the K–12 educator's work environment. Thus, through inquiry- and designed-based collaborations and increased access to the global infosphere, researchers, students and teachers can create and make available significant and meaningful outcomes and products of interest to themselves and to others. In the process they learn science concepts and ideas as well as the real "scientific methods" of inquiry and solving problems.

A LOOK AT SSPs AND TECHNOLOGY

If we investigate known partnerships to understand what is happening in the participating students' environments, in the teachers' rooms (or homes), and in the scientists' laboratories and offices, we find that successful partnerships are characterized by a number of common factors. These include the constant use of the Internet and access to rich existing scientific databases. New and friendly software developed by scientists and science educators to support collaborations allows analysis of data specific to areas of inquiry. Some software supports the initial need to scaffold students' understanding of the scientific concepts involved, and to facilitate discussions among the scientists and students. An expanding area of Internet use is as the electronic medium for publishing student research, for example, in on-line student research journals. These activities mirror the sharing of information between research laboratories and help the teachers become an integral part of the research program. Such integration leads to programs that may last years in a classroom, with one class of students moving on and the teacher becoming the research group's memory, keeper of the data, and facilitator of new questions to be researched by the new students. Exploring the science connections in the surrounding neighborhood with authentic problems in mind has been shown to provide ways to communicate the concept of inquiry-based activities to parents and the wider community, with subsequent increased support for schools.

In school-based partnerships, the keystone is the teacher. In some cases, the teacher is herself or himself the scientist, trained in a discipline, keeping up to date in the latest literature on the respective subjects and prepared to ask meaningful questions. Many of these scientist-teachers come from a discipline that depends on the use of technology, and others have become proficient in the technologies on their own or through innovative in-service courses. In other cases, teachers are the pedagogical experts who form partnerships with scientists for their own specific inquiry-based activities [for reviews of various partnerships, see Sussman (1993), Rock and Lauten (1996), and *The Role of Scientists in the Professional Development of Science Teachers*, NRC (1996b)].

A recently documented paradigm of SSPs, characterized by producing benefits for *both* students and scientists, was highlighted at a conference and public forum at the National Academy of Sciences in Washington, D.C. sponsored by TERC and the Concord Consortium, and funded by the NSF. This paradigm includes student observations and gathering of data that, either by themselves or in conjunction with other research data, answer questions of interest to the scientist and the scientific community. Such projects benefit the students by improving their understanding of the meaning of scientific data and their interpretation, by their participation in solving real and relevant questions, and by focusing their science learning on specific concepts in context. Examples of programs that were highlighted at the conference involve sequencing portions of genes as part of the Human Genome Project, recording the migratory patterns of birds and butterflies, gathering weather and earthquake data, making global environmental measurements, finding supernovas, participating in processing data gathered by the Mars Exploration, and monitoring ozone measurements (National Conference on Student and Scientist Partnerships, 1997).

For the most part, these partnerships are enabled by the use of different technologies in the classroom. Teachers' access to and comfort with the technologies used are important to the success of integrating SSP activities as part of the K–12 curriculum. Teachers as facilitators, supported by ongoing partnerships with one or more scientists, and teachers as providers of resources for students should lead to a rich environment where both teachers and students become learners. These partnerships will help to humanize the technologies and cast the partnerships as mutually meaningful relationships and collaborations. The Standards-based inquiry activities, especially when the students begin to ask their own questions, are well served by computers with Internet access to e-mail and databases, by databases and information on CD-ROMs, by video camera footage for gathering data, by special technologies such as Geographic Information Systems (GIS) for recording geographic data points, and by research tools that are relatively simple and inexpensive for measurements, to name a few technological applications. Electronic telecommunications that allow student–student, and student–scientist discussions of data, so well pioneered in such programs as the National Geographic KidsNet program, are still in use, complemented by Internet-based videoconferencing tools such as CU-SeeMe. Innovative software packages, such as Alice Network Software and Testbed for Telecollaboration by TERC, have served students in the analysis of data, in modeling the data and in preparing presentations for parents and the community. But more complex and easier-to-use software is needed to facilitate the partnerships; many of the current activities represent only the first generation. Many of these projects have been documented among other awards in NSF publications.

The involvement of scientists in these partnerships is only in its infancy. It would be helpful if credit for time away is allowed in the scientists' and engineers' workplace or university settings or other incentives were in place. However, scientists continue to involve themselves as the needs for science, technology, engineering, and mathematics outreach to the community become more and more apparent. Projects that bring together scientists and teachers to develop close partnerships need long-term commitments on the parts of both partners and must exemplify an atmosphere of trust and respect for each other's constraints, knowledge, and expertise. Although it is helpful and motivating for students and teachers to visit the laboratories of scientists and the workplaces of engineers, it is more valuable for the scientist to understand the school constraints and how to function within the environment of teachers and students. Scientists are often able to contribute the scientific materials necessary for experiments and may be able to provide guidance resources and help in developing the hardware and software for telecommunications and access to data resources. But there is a need to develop materials to guide scientists on what to expect when entering into these partnerships. At a minimum, scientists should have knowledge about what is in the National Science Education Standards, and have some familiarity with the programs that NSF supports in the K–12 community.

There are other concerns and questions related to the implementation of SSPs that need to be addressed in educational research. How are we assessing the learning that takes place in these complex interactions? How is the impact of student attitudes reflected in career choices or summer jobs? Can we document successes and failures of programs that use research and communication technologies in the schools? Do we understand how—and what—science children learn from these interactions in contrast to the more traditional, textbook-based lecture-based type of classroom? A recent book by Means and Olson (1995) focuses on case histories of uses of educational technologies in the classroom and suggests we need to understand not only how to measure their effects on student learning, but also how the way in which the learning activities are implemented influences student learning outcomes. Moreover, we must be aware of the many factors that may affect the success of that implementation. Little research has been done on the kinds of technology-supported, constructivist, student-based activities that are involved in SSPs, or what influence they have for long term impacts on these various communities.

THE ROLE OF RESEARCH AND NSF SUPPORT

The model of "student learning" implicit in SSPs differs from the one traditionally associated with teacher presentation of materials to a classroom audience, and to some degree also from "constructivist" models of learning. Thus, a discussion of the research that led to increasingly successful SSP implementations is instructive. As we explore the educational research associated with such interactions, we know that there is not yet a generally accepted and adequate means for experts to communicate dynamical, complex, abstract ideas to nonexperts.

A major paradigm shift in educational and cognitive science research, from theories of "learning" to theories of "cognition," is at the core of much of current educational thinking and practice. Cognitive science addresses how the human thinks and uses information. The emphasis is placed on developing thinking and problem-solving skills—for example, as practiced by experts in various domain fields—rather than focusing on teaching and learning based on disciplinary taxonomies. In the context of educational reform, there is a perceivable shift in educational goals from "understanding" to "knowing what to do" to reach a desirable outcome. There is a shift from accepting formulated problems to learning how to formulate problems and asking the right questions to solve them. All of these concepts are embedded in the kinds of inquiry-based activities that characterize SSPs.

Scientists and students alike acquire new knowledge incrementally, building on the knowledge they already have. The conceptual distance between prior and new knowledge—the size of the incremental step that can take place—depends on the methods and tools that can support it. Mathematical models and data or process visualization play different roles in the scientist tool kit. A large conceptual step could become easier for students with a pedagogy such as reflected in inquiry science in these partnerships, which might highlight conceptual manipulation and exploration of the model of a complex system while hiding many procedural steps (Jackson *et al.*, 1994).

What computer power and methodologies, such as visualization, have contributed to research is the ability to simulate more complex phenomena and explore the limits of the simulation visually. In areas of science where modeling and microscopic theories are important, simulation and modeling-based pedagogy can focus on conceptual advances while procedural manipulations can be performed and visually represented—reified—by a computer. Many of the advances in our ability to explore and exploit the resulting pedagogy are based on research and development work that has been supported by the NSF for many years. The aims of the Applications of Advanced Technologies (AAT, 1984 to 1996) program were to lay the conceptual foundations for advancing knowledge on how to use evolving technologies in support of education goals. This included exploring new scientific and pedagogical paradigms and methods, improving understanding of learning needs, prototyping technology-based science and mathematics programs, assessing the strengths and weaknesses of innovative applications, and measuring their effectiveness on performance.

The areas where AAT has created a significant educational knowledge base include effective technology-based science and mathematics tools that incorporate integrative methodologies of current science; interactive tools and technology-based learning environments, and new paradigms for distributed and collaborative learning and teaching through innovative uses of telecommunications. Many AAT projects have made their mark in education practice and thinking (LOGO for example) and include projects dealing with the interdisciplinary nature of chaos, fractals, and nonlinear systems; with the use of student-driven simulations, global change research, visualization, and computer modeling; and with students' contributions to and use of research data.

Specific examples of the accumulated growth enabled by sustained NSF research, which led directly to the SSPs discussed in this volume, are the development and prototyping of telecommunications projects such as KidsNet and the development of microcomputer-based laboratories—two research and prototype projects that led also to the creation of new market opportunities. Research and development projects that focus on formulating problems and solutions broadly implementable at a future time, such as the two mentioned, require a long lead time before their impact is fully known. It is also unlikely that such impact will be the result of a single project; it is the advancement of ideas and skills of the community of researchers and practitioners working and thinking in classroom situations that can impact the thinking of the education community. The increasing visibility and success of the partnerships highlighted in this volume and forthcoming publications are an integral part of this process initiated by AAT-funded research.

The invention and use of LOGO; a worldwide global laboratory for environmental measurements with inexpensive instrumentation; the integration of school-based seismic data gathering with earth science and physics education and with scientific research and professional databases; and many other examples attest to the contributions and long-term vision of research in the applications of technology to science and mathematics education. A look at the current projects supported by AAT is available via the Web server at http://www.ehr.nsf.gov.

There is increasing need for integration of technology into all aspects of education. Thus, the goals of the AAT program have been integrated into a larger, more encompassing one, called Research on Education, Policy and Practice (REPP), and are shared by the multidirectorate Collaborative Research on Learning Technologies activity, part of a new initiative of the NSF initiative titled Learning and Intelligent Systems (LIS). LIS will support multidisciplinary research that can create the next wave of rapid and radical advances in our understanding of and support for learning, creativity, and productivity—in short, advancing learning about learning. One of the goals of LIS is the nurturing of a strong education research community that can influence the path of research and development on learning technologies, and hasten our understanding of their application for education.

Through LIS, interdisciplinary teams will try to develop a deeper and more coherent understanding of how learning can happens in humans, animals, and in artificial systems. Supported research will try to understand what kinds of knowledge or skills can actually be learned in what kinds of environments, by what kinds of systems; how human, other natural, and artificial learning systems learn; what kinds of learning tasks are best suited for each; what natural processes and learning technologies best support each; and what kinds of knowledge each characteristically produces. Using this knowledge, researchers will explore how to develop a comprehensive collection of learning tools and methods that integrate linguistic, behavioral, biological, cognitive, and educational ideas with the development of interactive, collaborative, and multisensory technologies. These tools and technologies should be accessible to users with varied abilities and expectations.

SUMMARY

SSPs are an outcome of a successful research and innovation process. With the advent of pervasive communication technologies and more flexible and less expensive research technologies, these partnerships have the potential to enable a major change in the way science is taught in the schools. Innovations supported by various NSF programs have helped created this paradigm and demonstrate its power. But there is much left to be done. Various existing partnerships suggest a vision of classrooms in a mode of intense work—much of it self-directed—and of collaboration and learning—much of it peer to peer—on the part of the students as they participate in science activities and problem solving in a meaningful context. In this vision, the teacher would play a pivotal role as facilitator and coach for such activities and will use different measures to assess the knowledge of her students. The teacher would be computer literate, will integrate the use of technology as an important tool at her disposal, and will be able to seek resources through the Internet from partner scientists as well as from other teachers, from databases and other resources to support the student research activities. Outreach activities to communicate the students' constructed knowledge would be presented to the immediate or distant community, via symposia and on-line publishing. Nodes of school-based science activity, consisting of larger and larger groups of students, familiar with various research protocols and computer activities, will provide information about questions that were never considered by scientists—because of either lack of time or lack of local interest or knowledge. Students and scientists will become partners in creating knowledge that is new and useful. New understandings about how students learn and how teachers assess their learning with educational technologies will be generated through integrated and interdisciplinary research. These understandings will form the basis of innovative technological advances, giving K–12 students opportunities to benefit from the leading edge of the scientific research enterprise, and contribute to their own society's well being. Various programs at NSF are poised to receive proposals that foster innovations from scientists, teachers, and science educators, as individuals or in collaborative groups in support of these goals.

ACKNOWLEDGMENTS

ESIE Divisional accounts provide access to examples of GIS programs (NSF95-124), further information about Instructional Materials Development Awards (NSF95-124), and recent Teacher Enhancement Awards (NSF96-18). The REC Division makes available through its Web server (http://www.ehr.nsf.gov) summaries of the AAT awards, with links to project home pages. Information on the availability of these publications can be sought from <info@nsf.gov>.

REFERENCES

American Association for the Advancement of Science, *Benchmarks for Science Literacy. Project 2061* (New York: Oxford University Press, 1993).

B. Berenfeld, "Linking Students to the Infosphere," *Technological Horizons in Education Journal* 23(1996):76–83.

A. Collins, *National Science Education Standards*: Qualities and Quandaries. in R. W. Bybee, ed. *National Standards & the Science Curriculum: Challenges, Opportunities & Recommendations* (Dubuque, IA: Biological Sciences Curriculum Study. Kendall/Hunt Publishing Co., 1996)

Consortium for Educational Telecomputing: Conference Proceedings, R. F. Tinker and P. M. Kapisovsky, eds. (Cambridge, MA: TERC, 1992).

A. Feldman and H. McWilliams, *Planning Guide for Network Science* (Cambridge, MA: TERC, 1995).

S. Follansbee, *The Role of Online Communications in Schools: A National Study*, (Washington, DC: Center for Applied Special Technology, 1996).

K. Fulton (ed.), J. D. Wasser, A. Rubin, C. M. Grant, M. McConachie, A. Feldman, W. Spitzer, E. McNamara, and B. Porter, *Technology Infusion and School Change: Perspectives and Practices. Model Schools Partnership* (Cambridge, MA: TERC, 1996).

P. Horwitz and W. Barowy, "Designing and Using Open-Ended Software to Promote Conceptual Change," *Journal of Science Education and Technology* 3(1994):161–185.

S. L. Jackson, S. J. Stratford, J. Krajcik, and E. Soloway, "Making Dynamic Modeling Accessible to Precollege Science Students," *Interactive Learning Environments* 4(3)(1994):223–257.

D. H. Jonassen and T. Reeves, "Learning *with* Technology: Using Computers as Cognitive Tools," in D. H. Jonassen, ed., *Handbook of Research on Educational Communications and Technology* (New York: Scholastic Press and Association for Educational Communications and Technology, 1995).

C. L. Julyan, and M. S. Wiske (eds), *Learning along Electronic Paths: Journeys with the NGS Kids Network.* (Cambridge, MA: TERC, 1994).

R. Marshall, *Beyond National Standards and Goals: Excellence in Mathematics and Science Education K–16*, NSF 93-70 (Washington, DC: National Science Foundation, 1993)

B. Means, and K. Olson, *Technology's Role in Education Reform. Findings from a National Study of Innovating Schools*, ED Contract No. RR 91172010 (Washington, DC: U.S. Department of Education, 1995).

National Conference on Student and Scientist Partnerships, Conference. (Cambridge, MA: TERC, 1994 and The Concord Consortium, 1997).

National Research Council, *National Science Education Standards* (Washington, DC: National Academy Press, 1996a).

National Research Council, *The Role of Scientists in the Professional Development of Science Teachers* (Washington, DC: National Academy Press, 1996b).

NSF 95-24: NSF in a Changing World: The National Science Foundation's Strategic Plan, (Arlington, VA: National Science Foundation Publication, 1995).

NSF 97-18: Research in Learning and Intelligent Systems Program Announcement, (Arlington, VA: National Science Foundation Publication, 1997).

S. A. Raizen, P. Sellwood, R. D. Todd, and M. Vickers, *Technology Education in the Classroom: Understanding the Designed World* (San Francisco: Jossey-Bass, 1995).

B. N. Rock and G. Lauten, "K–12th Grade Students as Active Contributors to Research Investigations," *Journal of Science Education and Technology* 5(4)(1996):255–266.

N. H. Sabelli, "Essay: On using Technology for Understanding Science," *Interactive Learning Environments* 4(3)(1994):195–198.

Sussman, A. (ed), *Science Education Partnerships: Manual for Scientists and K–12 Teachers* (San Francisco: University of California Press, 1993).

U.S. Congress Office of Technology Assessment, *Education and Technology: Future Visions.* OTA-BP-EHR-169. (Washington, DC: U.S. Government Printing Office, 1995a).

U.S. Congress Office of Technology Assessment, *Teachers and Technology: Making the Connection.* OTA-EHR-616. (Washington, DC: U.S. Government Printing Office, 1995b).

S. Weir, "Electronic Communities of Learners: Fact or Fiction," in R. Tinker and P. Kapisovsky, eds., *Prospects for Educational Telecomputing: Selected Readings* (Cambridge, MA: TERC, 1992).

CHAPTER 11

Science Education as Driver of Cyberspace Technology Development

Roy D. Pea, Louis M. Gomez, Daniel C. Edelson,
Barry J. Fishman, Douglas N. Gordin, and
D. Kevin O'Neill

INTRODUCTION

Educational applications of networking technologies are becoming increasingly prevalent (National Center for Education Statistics, 1996; Riley et al., 1996). But "applications" are too often treated as infusions of technology into society, not drivers of new technological or research developments. One premise of the Learning through Collaborative Visualization (CoVis) Project challenges that common belief (Pea, 1993). Extending media-rich and highly interactive learning and teaching activities beyond single classrooms makes demanding requirements for new applications. We set out to create "distributed multimedia learning environments" to serve the emerging needs of precollege science education, which highlight learning through guided inquiry and affiliated new roles for teachers (National Research Council, 1996).

Distributed multimedia learning environments using advanced computing and communications technology can better enable precollege science education

Roy D. Pea, SRI International, 333 Ravenswood Avenue, Menlo Park, CA 94025; roypea@unix.sri.com. Louis M. Gomez, Daniel C. Edelson, Barry J. Fishman, Douglas N. Gordin, and D. Kevin O'Neill, Northwestern University, Evanston, IL 60204; l-gomez@nwu.edu, edelson@covis.nwu.edu, b-fishman@nwu.edu, gordin@covis.nwu.edu, and oneill@covis.nwu.edu to foster learning science by doing science, using the Internet. In our research, we have been investigating the requirements for placing such environments into classrooms and the challenges that must be met for their appropriation. The implementation and sustained use of such technologies on a large scale basis in K–12 school communities is a challenging design problem, related in large part to the complexities of sociotechnical systems.

The CoVis Project has provided a wide-band high-speed computer network and a desktop video conferencing network, structured groupware for collaboratively developing and conducting scientific inquiries, and scientific visualization tools that provide access to wide-ranging data sets on climate, weather, and other global parameters. The development of the CoVis learning environment and its components was guided by a question-centered and collaboration-focused pedagogy that recognizes students and teachers will continue the design process by innovating uses and suggesting revisions to their functionality.

During the 1995–96 school year, the CoVis Project involved over 100 teachers and more than 3000 students, using telecommunications as a critical leveraging technology to develop and evolve through use a wide array of new resources for learning. This testbed is now creating multiple opportunities for learners to engage in open-ended inquiry and new kinds of learning relationships. In essence, the CoVis Project has established a community of invention, which has spent the past 5 years constructing, experiencing, and evolving new kinds of science learning environments. This chapter provides a wide-angle view on the demands of science learning environments and on five CoVis examples of how these needs have led to new cyberspace technology developments.

THE CoVis PROJECT

Our goal has been to create a scalable approach to the establishment of distributed multimedia learning environments for high school and middle school science. We have designed and now provide wide-area broadband services that integrate educationally appropriate scientific visualization tools, newly developed structured hypermedia collaborative workgroup software, desktop videoconferencing and screen sharing, and standard Internet communication tools. Why did we take this approach?

Distributed Science Learning: Serving Science Education Needs with Cyberspace Technology Development

The CoVis Project was founded on the premise that classroom science learning should more closely resemble the open-ended, inquiry-based approach of science practice. As we put it in the January 1992 proposal to the National Science Foundation:

We believe that science learning environments should look and act more like the collaborative, connected work environments of scientists. To this end, teachers and students need ways to *reduce the complexity of getting access to resources* that are inaccessible locally. These resources include human expertise in the form of other teachers, scientists and graduate students in business, industry, and research settings, and other learners. They also include tools, instrumentation, hands-on materials and labs, museum exhibits, and computing and telecommunications infrastructures....

[We] aim: To examine how geographically dispersed teachers, students, and collaborators can *integrally and readily use* advanced information technologies to facilitate the types of collaboration and communication demanded by project-enhanced science learning.

We argue that applications of advanced technologies provide American educators with critical levers for promoting cognitive apprenticeships in science learning. We aim to build the next generation of infrastructure for new forms of science learning and teaching, and create a national model for the kinds of distributed multimedia science learning environments supportable with the future NREN.... We argue that these new [High Performance Computing and Communications] technologies can provide the backbone for the transitioning process from didactic science teaching to cognitive apprenticeships in project-enhanced science learning.... Students need to learn and do science in *context* of real problems and *with* sophisticated tools.

The CoVis Project thus designed a "network testbed" in which such a vision for science educational reform—that takes cognitive, social, technological, and scientific breakthroughs into account—could be concretely implemented and then empirically studied as a community appropriates and evolves its uses. We found it essential to attend to learning and teaching needs, as described below, in the context of technology trends. Our attention to trends in the evolving national information infrastructure when we began in early 1992 led us to client-server distributed network architecture, to scientific visualization as a substantive emphasis, to designs for new collaborative tools for enabling joint work among widely distributed CoVis community participants, and, initially to our use of a public-switched ISDN network. (We later moved to a variety of schemes for Internet connectivity as they became available.) In short, we sought to create a model community of *distributed science learning*.

Our assessment of learning and teaching needs led us to project-enhanced science learning as a fundamental pedagogy for achieving deeper understanding (e.g., Ruopp *et al.*, 1993). This approach to science education was developed as a combination of several current strands of research in the cognitive sciences of learning. Chief among these is the conception that, long before and also outside of formal education, people have learned through participation in communities of practice (e.g., Lave and Wenger, 1991). The advantage of such communities, in contrast to the regimens of "delivery-oriented" instruction common today, is that learning is situated with respect to community-based goals and activities in which knowledge is developed and used (Brown *et al.*, 1989; Pea, 1992; Songer, 1995). In the classroom, this advantage may take the form of what Collins *et al.* (1989) called cognitive apprenticeship, with students guided, both by their teachers and by remote mentors, to think about science in many of the fundamental ways that scientists do. In designing CoVis classroom activities, we quickly realized that there were vast distances, both cultural and physical, separating classrooms from scientists. The point of science education is not to take the "little scientist" that Piaget said was in every child and make him or her become an actual scientist. In recognition of this separation, our goal was to design settings in which students could become what Lave and Wenger (1991) have called "legitimate peripheral participants" in communities of science practice.

In the community of science practice, the tools used and the data collected for investigations are critical to the scientists' abilities to formulate, think about, and work on their problems (e.g., Cerf et al., 1993) . In designing the CoVis testbed, we sought to take advantage of this reliance by providing students with ways to access the same data the scientists study using tools modeled on those used by scientists. Providing access to these tools and the scientists and other professionals who use them is one of the key design techniques we used to foster a sense of legitimate peripheral participation. There are other recent efforts to bridge the science education–scientific practice "gap," including Global Lab, Project GLOBE, GREEN, EARTHWATCH, and Hands-On Universe, which have developed related models, of "student–scientist" partnerships (see this volume).

Education Is a Demanding Environment

Computer and communications science and industries make a mistake when they assume that advances in technology stem either from technological innovations and engineering alone, or from generalized theories of human–computer interaction. We know from work this past decade on "user-centered system design" for software systems that it is exceptionally important to fit the tool to the task, even as we seek to invent technology paradigms and applications so that new tasks become possible (e.g., Greenbaum and Kyng, 1991; Nardi, 1996; Norman and Draper, 1986; Schuler and Namioka, 1993).

Education in particular has suffered because it is often the last application area for advanced technologies. It has typically taken 15 years for computer tools to make their way into classrooms from their advent in military, industrial, and university settings (U.S. Congress Office of Technology Assessment, 1987). Examples include Logo programming, spreadsheets, graphing tools, simulations, and word processing. But education is far too crucial a function for the future of our increasingly global societies to wait in this long line.

In precollege education, children of different ages participate daily in one of the most demanding challenges of a lifetime. They are supposed to devote some 18,000 hours over about 200 seven-hour days a year for 13 years to learn to engage the different abilities of the mind and body, to become literate in the subject matter, skills, and media of their era, and to contribute to society in work, family, and community. Until recently, students' primary setting for doing so was with one or a few teachers, within a few small rooms, not connected to the world even by a telephone.

Education is not only an "application area," but what we will call a high-priority "invention area." Industry could benefit from working with educational professionals and researchers in the learning sciences to understand the market opportunities surrounding the provision of better support to the activities of learning and teaching.

Although teachers have adapted previous "innovations" such as slide projectors, audiorecorders, radio and TV without making substantive change in their teaching (Cuban, 1986; Tyack and Cuban, 1995), there is strong evidence that teachers are recognizing how computer and communications technologies have vast potential to provide new work tools for transforming their educational practices and the learning experiences and outcomes that are possible in fundamental and exciting ways (Fisher *et al.*, 1996; Means, 1994).

The demands and constraints of educational settings can drive the development of very interesting and valuable new technologies and paradigms of computer and telecommunications use in support of human activities. Several years ago, Pea and Gomez (1992) made this argument for the creation of distributed multimedia learning environments that integrate computing and communications to provide students with rich resources for learning accessed in the context of geographically separated learners.

Science Education as Driver of CoVis Technology Developments

We now provide five examples, in a common framework, for how the demanding needs of science education have led to new cyberspace technology developments in the CoVis Project. A preliminary overview of the CoVis network environment and "testbed" suite of applications will be useful. We began working with six earth and environmental science teachers at two Chicago-area high schools in summer 1992. In fall 1993, these teachers and nearly 300 students began the school year with scientific visualization tools for atmospheric sciences and an asynchronous collaboration environment called the Collaboratory Notebook, both developed by the CoVis Project, as well as desktop video teleconferencing, and a full suite of Internet tools including e-mail, Usenet news, and Gopher. Students in each classroom used these applications running on a group of six Macintosh Quadra workstations connected to a high-speed video and data network. These applications collectively provide a "collaboratory" environment (Lederberg and Uncapher, 1989; Cerf *et al.*, 1993) that couples tools to support communication and collaboration with open-ended scientific inquiry tools. As of November 1996, we are now working with over 40 schools, more than 100 teachers, and over 3000 students who are participants in the CoVis Collaboratory testbed.

Network Infrastructures for Shared Media Spaces

Context of Problem. K–12 schools have not generally benefited from internetworking technologies. Although there were over 5.8 million computers in use in schools as of 1995, only 12% of schools at that time had connections to some kind of wide-area network (U.S. Congress Office of Technology Assessment, 1995), and fewer than half of those schools had connections to even one classroom (Heaviside *et al.*, 1996). As many as 50% of teachers have little or no experience at all with technology in the classroom (McKinsey & Co., 1995). As the June 1996 report to the nation from the Office of the U.S. Secretary of Education characterized the current situation:

> Computers and information technologies are not part of the way most American students learn. Today's students spend an average of only a few minutes a day using computers for learning. Only 4 percent of schools have a computer for every five students—a ratio sufficient to allow regular use. Only 9 percent of classrooms have connections to the Internet. (Riley *et al.*, 1996)

One major goal of the CoVis Project, which began in 1992 when school internetworking was far less developed than in its 1996 state just described, has been to combine prototype and off-the-shelf applications to create a reliable, networked environment that showcases High Performance Computing and Communications technologies for K–12 learning communities. We intended to circumvent the traditional decade-long "trickle down" of technology from industry to education (Pea and Gomez, 1992) and design a first-rate suite of technology to support new forms of science education.

Educational Needs of Learners and Teachers. As in related industrial work at Bellcore and Xerox PARC, we wished to create rich "media spaces" (Gomez *et al.*, 1994) to support collaborative work and learning among adolescent students, their teachers, and scientist mentors. Research in the learning sciences has indicated that learners and teachers need highly interactive conversational environments around media-rich artifacts to provide common grounds for fostering learning communications (Pea, 1994; Pea and Gomez, 1992).

What Did We Develop? Although our goals have not changed in the 5 years CoVis has been working in schools, our approach to school networking and desktop videoconferencing has changed substantially from the early implementation of CoVis in just two schools to our present day implementation in 43 schools.

In the first phase of CoVis, the network design and implementation was the result of a collaboration between Northwestern, Ameritech, and Bellcore. We selected public-switched Primary Rate Integrated Services Data Network (PRI-ISDN) as the transport layer for the CoVis network. ISDN bandwidth can be broken up into call channels, and dedicated to different functions, so we were able to create a hybrid, two-function "overlay" network that gives student workstations access to both Ethernet-based packet-switched data services and circuit-switched desktop audio/video conferencing. Each CoVis classroom in the first 2 years of the project contained six networked Macintosh workstations with an accompanying desktop video teleconferencing unit.

At each school, the local Ethernet was bridged through an Ethernet-ISDN bridge and passed through an inverse multiplexer across the ISDN lines to Northwestern's campuswide network. We sought to avoid video call blocking, so each school had enough line capacity to maintain six video calls at once, while still providing T1-equivalent bandwidth for Internet connectivity. The initial CoVis network layout is shown in Fig. 1.

Desktop video teleconferencing is a critical component of the CoVis testbed. In the first 2 years of the CoVis Project, students used Bellcore's Cruiser application (Fish *et al.*, 1993), allowing them to place point-to-point video teleconferencing calls to other CoVis addressees by selecting an individual's name from a directory. Cruiser is a client application of Touring Machine (Bellcore Information Networking Research Laboratory, 1993), the distributed network software developed by Bellcore for managing heterogeneous media resources (e.g., cameras, microphones, monitors, switch ports, directory serv-

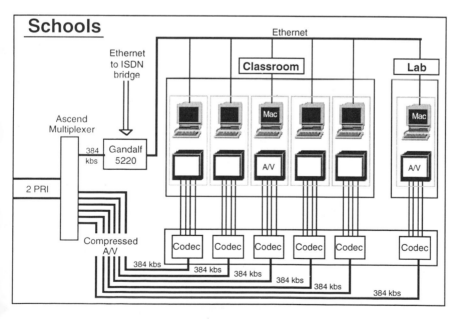

Figure 1. The network configuration for the two Year One CoVis schools, showing both data and video networking.

ices). It is significant that CoVis Project needs and affiliated Ameritech interests drove the first integration by Bellcore of Touring Machine into an ISDN network. Bellcore's Touring Machine and Cruiser development teams devoted over 2 person-years to the CoVis Project to develop network infrastructure. To our knowledge, CoVis was the first school-based application of ISDN desktop videoconferencing.

In addition to videoconferencing and Internet applications, the CoVis communications suite includes screen sharing. CoVis participants may collaborate synchronously through screen sharing, using Timbuktu (from Farallon Computing), which enables one user to see exactly what appears on the screen of another user, regardless of distance, and users at either end of a screen-sharing session may control the remote computer with their own keyboard and mouse.

In the second phase of the CoVis Project, we did not attempt to devise custom network infrastructures for each participating school. With 41 new schools joining the project, this was not feasible. However, it was also unnecessary, for during the first phase of the CoVis Project a new industry in Internet service providers (ISPs) was quickly becoming established. At the start of the second phase of the CoVis project, it was possible to ask schools to make their own arrangements for networking, at either 56-kb or T1 levels of connectivity. Instead of network implementers, the CoVis Project served as consultants to these schools, advising them on how to acquire the highest bandwidth at the best price, and how to think about the design of local area networks within their buildings.

During the time that CoVis was experimenting with Cruiser, a new form of videoconferencing tool was starting to emerge that used TCP/IP, or Internet protocols, as its transport mechanism. The premier example of this type of tool was CU-SeeMe, from Cornell University. The primary technical advantage of this software was that it enabled extremely low-cost videoconferencing—a school's investment in Internet connectivity also provided their video network. The primary technical limitations for this software in its early stages were a lack of integrated video and audio, and the sometimes complicated process of introducing video into one's computer system. Also, because of the variable bandwidth limitations of the Internet, video quality could be quite low. These tools are "packet-switched" (as opposed to "circuit-switched"), which means that they must share bandwidth with all of the other services in use on the network. Average frame rates for CU-SeeMe were between 5 and 8 frames per second of black and white video, compared with Cruiser's 25–30 frames per second of color video. And, at least initially, you needed to use a telephone to provide audio.

In early 1994, two technological developments made it possible for the CoVis Project to consider CU-SeeMe as a replacement for Cruiser. The first of these was the introduction of synchronized audio in the CU-SeeMe software, using the computer's built-in microphone as the source. The second was the introduction of the Connectix QuickCam, a low-cost digital camera (widely known as the "eyeball" camera for its shape) that required no additional hardware to get video into the computer. These new developments meant that most CoVis schools could now afford to participate in videoconferencing activities, and we discontinued use of Cruiser in favor of CU-SeeMe.

How Did It Work in Use? Our key result from phase one of the CoVis Project was that we were able to establish our ISDN-based network and put it in daily use by approximately 300 people, mainly high school students. The challenge of this effort was to take a collection of technologies, many only demonstrated or tested in small-scale lab situations, and place them into daily service in demanding conditions. Our progress culminated in a stage-by-stage installation in fall 1993 of the CoVis network testbed using public-switched ISDN services. It was used in the schools for approximately 2 years, until it was replaced by more conventional T1 lines and IP-based videoconferencing using CU-SeeMe (see Desktop Video section for details). During the first year of full implementation, we encountered a range of difficulties stemming from the experimental nature of some of the software, inexperience with novel products and services both within the project and at the commercial service providers, and product and network unreliability. Within 3 months of its initial deployment, however, the data and video network were stable and reliable. How specific applications were used within this networking environment (e.g., visualization, groupware, desktop videoconferencing) will be described in the sections below.

As CoVis moved into its second phase in late 1994, the challenges posed by network development were quite different than in the first phase. Of the 41 new schools joining the CoVis Project at that time, only 5 had prior Internet access. The remaining schools were interested in CoVis in part because we would assist them in obtaining Internet access, and in part because we provided schools with a substantial curricular context for justifying Internet access. The 36 schools that required Internet access all planned to obtain networks during the 1994–95 school year. However, only about half of these schools were actually able to complete their networks during this time period. There were two types of difficulty: First, many schools encountered political difficulty in obtaining permission to install networks from local school boards, related in part to concerns about pornography and the Internet raised by popular press articles during the summer of 1995. A second difficulty was unanticipated physical barriers within school buildings, including asbestos problems, outdated electrical wiring, and lack of adequate security for computers and network hardware. Finally, many schools simply had inadequate technical staff for dealing with unforeseen implementation issues that CoVis staff were unable to resolve from a distance. These issues were particularly challenging for the many urban schools we are working with in this scaled-up testbed, which were among the last to gain network access. All CoVis schools were finally wired by the Fall of 1996, nearly two full years later than many anticipated gaining network access. Among the

43 CoVis schools, 18 have established Internet connectivity with dedicated 56-kb networks, and the remaining 25 have dedicated T1 networks.

What Does This Tell Us about Design Both Generally and Specifically? CoVis experiments with ISDN networks during phase one taught us valuable lessons about the complexities of bringing networks into schools. Although the specific configuration of our ISDN network was not carried through into the second phase of the project, the general lesson to "expect the unexpected" in terms of system start-up problems remained relevant. Among the most important resources for schools to have is reliable technical support *in the building*. This is necessary not only for the design of networks, but also for their continued maintenance.

What Challenges Remain? Although we now have 43 schools up and running as part of the CoVis Collaboratory testbed, it is still not routine for schools to have Internet networks installed. This is the result of a mixture of technical and social issues. Schools need budgetary support to hire qualified network specialists. Internet providers need to devise pricing structures for school Internet connectivity that fit within the complex budgetary structures of schools. And projects like CoVis need to continue developing compelling activity frameworks to make these Internet connections worthwhile investments for schools.

Adapting Scientific Visualization Tools for Learners

Context of Problem. When we started our work on scientific visualization for precollege education in early 1992, we were impressed with the utility of general-purpose visualization environments for the scientific research community. Scientists were using visualization to find patterns in complex data using the distinctive strengths of the human visual system (Brodie *et al.*, 1992; Wolff and Yaeger, 1993). Demonstrating support for this method of research were the standardization of data formats and distribution of visualization packages. Science teachers and students could benefit from similar means if they were helped to appropriate and "redesign" visualization for their needs (Gordin and Pea, 1995).

In studying science, students are commonly confused by the abstractions, formalisms, and quantitative terms of equation-based data representations. By taking advantage of the powerful capabilities of human visual perception, scientific visualization offers a new route to scientific understanding, and the possibility of reaching students traditionally ill-served. Scientific visualization also offers the possibility of opening up new domains for study that have been considered too complex for students because of their heavy reliance on formulas and abstract representations. Similarly, scientific visualizations can give students the ability to conduct data-driven investigations in areas to which they previously only had access to broad overviews (e.g., global climate data; planetary biomass distribution; ocean temperature).

We had spent considerable time in science classrooms, and knew from previous work on science educational software that a vast gap existed between what students understood and what they would need to know to use such tools effectively in support of science learning.

Educational Needs of Learners and Teachers. Substantial revisions were necessary to make scientific visualization technologies usable for precollege science students (Gordin and Pea, 1995). Typically, scientific visualization software, such as Spyglass's *Transform*, provide powerful general purpose means to render data as images. For example, *Transform* can be used to render temperature and precipitation values as a colored image where the temperature is shown through color and precipitation through contours. The data sets for scientific inquiries like these are often so well known by the scientists who use them that they do not even bother to label the variables represented. The operations in these visualization systems are extensive, as they are designed to cover the common operations of many domains. Scientists typically do not even notice all of the irrelevant or nonsensical operations that are allowed (e.g., subtracting temperature from precipitation). Instead, their actions are guided by the underlying semantics of the quantities they are manipulating that lead them to enact well structured idioms of operations. In this way, scientists demonstrate shared tacit knowledge about the data sets that students lack. For example, scientists are keenly aware of the distortions that spatial projections introduce and factor them into their analyses. Yet students are often unaware of the misleading properties of these projections nor how to render the data so as to compensate for their effects.

What Did We Develop? To adapt scientific visualization for the classroom we "knowledge engineered" atmospheric scientists and reified their analysis techniques into our new visualization tools for learners. We sought to rapidly prototype these new learning environments by building interfaces "in front" of existing scientific visualization tools, thereby creating learner-appropriate interfaces to these tools' already impressive functionalities. These "front-ends" provided supports for novices to learn the techniques and idioms of visualization in a structured way. We defined a four-step methodology for adapting the scientists' visualization tools for education:

1. *Investigate science practice.* Scientists are observed using visualization tools and data sets for a specific domain, with the goal of eliciting the sorts of questions the visualization tools and data sets can be used to investigate for their area, and how the tools are employed during

inquiry. A crucial step is choosing a set of highly interrelated data sets within which students can investigate relationships, thereby promoting the development of causal accounts on their part for the domains under study.

2. *Identify tacit knowledge used in science practice and make it explicit within our visualization tools.* We seek to articulate the tacit knowledge scientists employ when using visualization tools, e.g., scientific principles, understanding of the limitations of data resulting from collection processes and underlying models, and common means used to probe the data for patterns or anomalies. This knowledge is then made explicit in the software interface to assist students in pursuing meaningful questions.

3. *Motivate visualization in terms of students' interests, experiences, and conceptions.* For example, we often begin students' use of visualization by asking them to draw a visualization of any quantity they choose. The often striking drawings students produce demonstrate visualization as a new expressive medium that students have used (see Fig. 2) to render widely ranging quantities (e.g., unemployment rate, biomes, and use of video games). Similarly, we have designed activities that focus on phenomena familiar to students. For example, using visualizations of incoming sunlight and temperature we problematize the changing of the seasons by identifying apparent anomalies in the observed patterns. Through these steps, we intend visualization to become a tool ready-to-hand for students to investigate problems that matter to them.

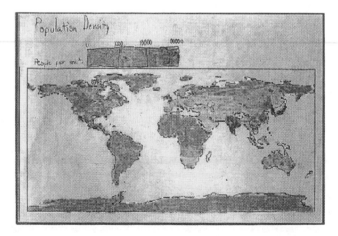

Figure 2. Student hand-drawn visualization showing global population.

4. *Refine the visualization tools and activities in response to formative evaluations.* Through a combination of observation and direct user feedback, we evaluate the patterns of use that emerge and iteratively redesign the software and activities as needed.

We developed four visualization environments using this process. These visualization environments primarily focus on atmospheric data, although the last also includes human geographic and environmental data sets. These visualization environments are called:

1. *Climate Visualizer* (Gordin et al., 1994; Gordin & Pea, 1995)
2. *Weather Visualizer* (Fishman and D'Amico, 1994; Ramamurthy et al., 1994, 1996)
3. *Greenhouse Effect Visualizer* (Gordin et al., 1995)
4. *Climate Watcher* (Gordin et al., 1996a,c)

The first two of these are built on top of a scientific visualization tool used by researchers (e.g., *Transform* from Spyglass, Inc. for the Climate Visualizer, and *WxMap*, at the University of Illinois, Urbana-Champaign, for the Weather Visualizer; Ramamurthy et al., 1994). The third uses the World Wide Web to provide visualizations and offers users the ability to query the values of specific points and inspect the visualizations at varied resolutions. The fourth is a stand-alone Macintosh application with a wide range of data sets and many means for rendering and creating visualizations. The increased development costs for this last visualizer were justified by the need to consolidate the findings of prior prototypes (especially the Climate Visualizer and Greenhouse Effect Visualizer). This process illustrates an important way that we have engaged in formative research. Initially rapid prototypes were crafted based on interviews with scientists and other experts, then these prototypes were distributed to teachers and students who aided us in refining the software and collaboratively constructing activities. Only after the design had been refined through use and the creation of productive activities were our designs set into more robust and stable software.

How Did It Work in Use? During 1993–94 and 1994–95, students in the classrooms of the CoVis testbed used the Climate Visualizer and Weather Visualizer in their science projects. Although they found the tools comprehensible and the visualizations useful to offer new insights into several scientific phenomena they were seldom used (e.g., McGee and Pea, 1994; McGee, 1996). This was primarily related to the lack of structured activities that would provide teachers and students with an understanding of their appropriate contexts for use. As a result, teachers perceived relatively narrow curriculum scopes in which these specific scientific visualization programs fit (McGee, 1995), and students often lacked the ability to develop meaningful uses of these investigative tools on their own.

To address this issue and the broader need to provide a common fulcrum on which to leverage the changes CoVis was bringing to bear, a set of global warming curriculum activities and resources was created and implemented in 24 schools in 1995–96 (Gordin *et al.*, 1996a,b). An integral part of this curriculum were visualization activities where students created visualizations on paper and on Styrofoam models, investigated the causes of the seasons, and analyzed the Earth's energy balance. These activities provided a huge impetus for schools to use visualization and many schools engaged in these activities. However, the style of use ranged widely. At some schools students genuinely engaged difficult issues of representation and meaning by investigating how the data had been collected and the causes of the visual patterns they observed. At other schools the visualizations were merely used as props by teachers to illustrate concepts. In these cases, visualizations became merely another set of color plates added to their textbooks.

What Does This Tell Us about Design Both Generally and Specifically? The iterative and collaborative nature of these designs, which moved technology from the scientists' laboratory to the classroom, requires rapid prototyping to accommodate the multiple cycles of formative research. We successfully built several prototypes by "front-ending" scientific tools with pre-college-age appropriate interfaces, thereby creating a high-end scientific workbench for them. This was an important achievement for us and for our industry partner Spyglass who foresaw a much vaster audience and market for their work—including children with little background in science—thereby opening them to an entirely unexpected market from their perspective.

During this formative process we came to view scientific visualization as a new classroom medium that must allow for student expression and interpretation. Initially, we provided teachers and students with scientists' data and representations and asked them to use them in ways that were analogous to scientists. Over time, we came instead to ask how teachers and students can "play" with visualizations, that is, craft new visualizations and assimilate the medium to their own interests and concerns. This shift does not mean we gave up on the utility of scientists' data or the importance of engaging students in the problems and practices of scientists. Rather, we saw the need for a developmental progression that moved through the stage where visualization is used to serve teachers and students existing concerns before it can serve as a vehicle to introduce new concerns. The need for this step was most palpable when we worked with younger students who had little patience with constructing verbal explanations for complex images, but were excited by the possibilities of crafting colorful images depicting their views of the world. In their images students often depicted human-centered quantities (e.g., population) leading us to add human geographic data to our visualizers and to realize that investigating human spatial patterns can give students a powerful steer on the road toward visualization.

What Challenges Remain? Our success in introducing visualization to science education rests largely on our teachers seeing advantages in using it. In particular, major challenges lie in adapting teacher pedagogy and curricular scope. When teaching is defined in terms of didactic lectures and structured laboratories, there is little room for the open-ended inquiry we are providing tools to support. Similarly, when the curriculum is scoped as covering a long list of topics listed in a textbook's table of contents, there is little room for project study. We believe that "weather" and "climate" and "greenhouse effect" need not be set topics that last for a set number of weeks in the earth sciences curriculum, but topics that involve such interesting, diverse, and complex phenomena that they could be investigated in projects throughout the school year in diverse classes. But to do so required too large a change in teaching philosophy and curriculum theory for most CoVis teachers. On the other hand, it would not have been practical for the CoVis team to conduct the ethnographic observations and interviews that were integral to the design of our knowledge-scaffolding front-ends to the visualization software and data sets for each and every one of the other cognate fields—astronomy, geology, oceanography—which the teachers sought to "cover" in their teaching of earth systems science.

An additional challenge is the development of activities that motivate teachers' initial usage of visualization. So far we have primarily studied how to ignite students' interests in visualization, especially by having them construct their own visualization and engaging problems with which they have experience. Similarly, we must better understand how to meet teachers' existing goals with our visualization environments.

A Groupware Application for Systematic Science Inquiry

Context of Problem. Early approaches to the use of technology in education were based on a transmission model of instruction, in which the technology (e.g., film and broadcast media) was used to transmit instruction in a more engaging fashion and to larger numbers of students. Distance education inherits this tradition when it uses phone lines, satellite links, and microwave to transmit static knowledge to wider audiences, with minimal opportunities for highly interactive conversations with instructors or learners (Pea and Gomez, 1992).

With the advent of the personal computer, technology took on new roles for learning, including providing responsive environments such as those described in the previous section that support active learning through investigation. However, in the workplace and other environments, computers have also shown great value in supporting communication and collaboration, as we see with e-mail and in groupware tools such as Lotus Notes. Increasingly, technology has assisted in broadening the form that collaboration takes to include not just discussion, but the sharing of artifacts and cooperative work across time and distance in shared "media spaces." Technologies with similar emphases can play a revolutionary role in supporting new forms of learning conversations in educational settings (Edelson et al., 1995b; Pea, 1994).

It has become a commonplace for cognitively guided research in science education to document how ineffective lecture- and demonstration-centered pedagogy is at securing student's understanding of subject matter (Linn et al., 1995). Learning that builds from students' questions, and engages the knowledge that they have, is characterized as a "constructivist approach" to learning (e.g., Hawkins and Pea, 1987; Papert, 1994; Tobin, 1993). Projects in the sciences can build from students' questions, and when well guided by mentors, can provide motivating and effective contexts for the acquisition of research skills and scientific understanding (Ruopp et al., 1993). With this in mind, we set out to create a groupware application to foster student learning and systematic science inquiry through project-oriented learning.

Educational Needs of Learners and Teachers. When teachers seek to integrate open-ended science projects into classroom life, they find it difficult to maintain records of students' inquiries and monitor their progress on investigations. We expected that well-tailored software could provide significant support for these efforts, especially in a learning environment where many of the artifacts that students and teachers would want to share in the course of their work might be electronic.

What Did We Develop? To meet this challenge, we created a wide-area network-based hypermedia "Collaboratory Notebook." This software was built with an Oracle database server and TCP/IP client software as a foundation (Edelson and O'Neill, 1994), and many classroom-specific customizations on top. The software uses a unique set of inquiry-oriented hypermedia link types for all notebook entries. These link types were intended to "scaffold" students' processes of scientific inquiry, and enable distributed workgroups to more clearly define and conduct science projects in a stepwise (though iterative) fashion.

Within its hypermedia notebooks, the Collaboratory Notebook enables students to record text, tables, graphics, sound, video, and animations throughout their work. These products can then be seamlessly shared with others on the network. With these capabilities in hand, project groups may be composed of students within schools, across schools, and can involve adults as mentors, including teachers, scientists, science graduate students, and science education researchers. Project workgroups are enabled to negotiate project topics with one another and their teachers, and log observations from scientific visualization tools and hands-on investigations as they complete them. If students' records of their project work are reasonably complete, these can provide teachers and remote mentors with a valuable window into students' inquiry processes and activities.

In an example of how we envisioned the Collaboratory Notebook being used, a group of students would begin by developing an idea for an inquiry, negotiating its details with their teacher, and recording questions and hypotheses electronically. This would be followed by a plan for pursuing their inquiry. A teacher or other mentors located at a distance could, at their convenience, read the students' questions, hypotheses, and plans, and add comments to help them focus their efforts or to alert them to possible stumbling blocks or useful resources. Students would not need to be tied to one another, could go on to engage in separate research activities and individually record these for the others to view at a later time. Students could also electronically exchange questions and comments on their findings, draw conclusions, or initiate further research.

How Did It Work in Use? Beginning in the 1993–94 school year, teachers and students used the Collaboratory Notebook in a variety of ways to support experimental classroom activities. In one case, a teacher conducted a weather prediction activity using the Collaboratory Notebook, in which groups of students recorded predictions for the following day's weather and evidence to support those predictions. Students then had the opportunity to view and comment on the predictions of other groups within the Notebook software. The following day, the predictions were reexamined and discussed in the light of firsthand observations of the weather.

Other, less successful activities planned by CoVis teachers involved students recording questions and progress in the Collaboratory Notebook during more traditional classroom research projects. In these projects, students used few resources beyond standard textbooks and reference materials, and had no audience for their work besides their teacher and the class itself. These activities provided little motivation for students to use groupware, and they sometimes expressed their frustration at this to both the teachers and the researchers.

The Collaboratory Notebook was intended for an environment where access to computing resources was not expected to be a limiting factor. In fact, given access generally to only six workstations in each classroom of 25 students or more (and the slow rate at which many high school students type), limitations on computing resources became a serious constraint on curriculum planning. Accordingly, many projects were developed with only minimal use of the Collaboratory Notebook (Edelson et al., 1995), and some teachers elected not to use the software at all.

One reason why these outcomes are disappointing is that many projects conducted in classes throughout the CoVis testbed could have benefited from the mutual student-to-student influence that the Collaboratory Notebook supports. However, students in each classroom were generally unaware of one another's efforts (O'Neill et al., 1995), and cross school collaborations rarely transpired. One distant collaboration using the software occurred when CoVis teachers and staff arranged for mentors from the Atmospheric Sciences Department at the

University of Illinois and the Exploratorium Museum in San Francisco to advise students on their investigations. However, without this coordination effort, the cross-school work that was a key target of the software proved infrequent.

What Does This Tell Us about Design Generally and Specifically?
The scaffolding link types in the Collaboratory Notebook were quite attractive to some teachers. In these classrooms they were used to reinforce the teacher's guidance of students in their project work, and aided teachers in their efforts to monitor and shape student progress. Without a classroom culture to give them meaning, however, the link types seem to have had no magic of their own. Heeding the structure of the work tasks in classroom science projects, rather than using off-the-shelf software with untyped links, seems to have paid off for us; but only where teachers were ready to take advantage of and reinforce the software's design.

What Challenges Remain? We would like to see much more integral use of the Collaboratory Notebook in science learning in and out of the classroom. Such use would require several major changes to existing teaching and learning practice, however. One of these is the development of classroom cultures where presentation, commentary on, and revision of projects is more common. Our software is of the greatest use in supporting the revision of project work through helpful, critical commentary by partners distributed across space and/or time. Currently, however, most of the project work we observe is a one-time affair, and does not involve remote partners because of the management effort this requires. Some of our current research focuses on the development of network resources to ease the coordination effort associated with distant mentoring (O'Neill et al., 1996).

Another challenge to the more integral use of the Collaboratory Notebook relates to the many projects under way in the CoVis testbed which have relevance to one another, yet do not find their way into student and teacher discussions (on-line or otherwise). For example, tens of earthquake and hurricane projects are conducted every year by CoVis students, yet these rarely build on each other's findings. For our software to reach its maximum usefulness, this routine would have to change. This aim would require establishing a critical mass of classrooms throughout our network with students wishing to do projects together, and sufficient experience among their teachers to establish common standards for assessing the quality of the learning and research produced by these students.

A third and final change that would increase the use of software like ours would be the availability of a computer and network connection to every student and her teacher, both during school and, ideally, after school. The importance of making a computer and network connection available after school hours is especially great for teachers, for whom it is necessary to review student work at night and at home. An alternative in this vein would be to develop designs for groupware applications that would encourage use without each person having personal, on-demand access to the network.

Desktop Videoconferencing for Classroom Use

Context of Problem. Considerable enthusiasm has been developing in the industrial community, and in the "new media" marketplace for video-to-the-desktop. The industry focus has been on supporting formal videoconferencing participation by groups who do not need to go to a special facility, or on supporting the informal communications, much like running into a colleague in a coffee area, that tend to enhance the chances that such colleagues will engage in collaborative work (Fish *et al.*, 1993). There have been complex technological and social design issues. Technologies for desktop videoconferencing have been expensive primarily because of codec costs, and sustaining acceptable frame-rate performance for person–person communication needs over either LANs or WANs has provided a demanding environment for engineering. On the social side, the crafting of software and communication protocols for launching and managing desktop videoconference calls to support human communicative needs at work has taken considerable effort. When we began in early 1992, videoconferencing to the desktop in the classroom was virtually nonexistent, although many examples existed of "distance learning" uses of classroom-based videoconferencing (Pea and Gomez, 1992).

Educational Needs of Learners and Teachers. Our expectation was that there were a range of likely needs of the activities that go on in a project-enhanced science classroom which desktop videoconferencing could fulfill. First, we conjectured that a teacher's use of initial resource materials to motivate student interest in a science topic from which projects could be developed, such as videotapes, would carry over to a use of remote visits to interesting video sources, such as exhibits at the Exploratorium Science Museum (a CoVis partner). Second, we conjectured that students would wish to conduct some of their collaborative work across schools with the additional media channel of desktop video, to enhance the communicative feedback possible as they worked to further their project activities. And third, we conjectured that being able to create a distributed audience for the presentation and discussion of project results would be an appealing use of the desktop video medium.

What Did We Develop? We made it possible, with extensive engineering and redesign work from Bellcore and Ameritech described earlier, for students to use the Cruiser application to place calls to other CoVis addressees, including other classrooms and Northwestern researchers (see Fig. 3). Late in the 1993–94 school year, Cruiser installations at the Exploratorium Science Museum in San

Figure 3. High school students using Cruiser and screen sharing of visualizations.

Francisco and at the Department of Atmospheric Sciences at the University of Illinois, Urbana-Champaign made possible virtual field trips and interactive weather briefings.

How Did It Work in Use? In our initial discussions with teachers prior to the CoVis "launch" during the 1993–94 school year, Cruiser was viewed with great interest. This was understandable, as videoconferencing clearly offered the most immediately comprehensible and dramatic changes of all of the new CoVis technologies being introduced to their classrooms. Scientific visualization and collaborative groupware took longer to comprehend.

As researchers, we were also very interested in the success of Cruiser because it offered a real-time bridge between the communities of schools and scientists. Furthermore, Cruiser also required a sizable investment of funds to become operational, including ISDN telecommunications charges, hardware for video compression and decompression, cameras, monitors, and personnel to help set it up, debug it, and keep it running. Cruiser was technically available in CoVis classrooms beginning in January 1994.

Yet even with the extra emphasis placed on it in the overall design of CoVis classrooms, Cruiser was hardly ever used in daily classroom life. Tracking of the use of all computer-mediated computer tools during the 1994–95 academic year showed that whereas students sent over 10,000 e-mail messages, posted over 1400 Usenet news articles, and created over 1100 pages in the Collaboratory Notebook, there were only 32 Cruiser calls placed (Fishman, 1996). Surprisingly, it was used primarily for troubleshooting pedagogy and software use between teachers and Northwestern research staff, not between classrooms in cross-school collaborative projects as anticipated. How can the extremely low use of Cruiser be explained?

Many of the major difficulties were sociocultural in nature, not technical. Although teachers readily accepted the challenges of creating highly collaborative and interactive environments supported by asynchronous tools (e.g., e-mail, Usenet news, CoVis Collaboratory Notebook), they were more reluctant to include synchronous communication tools in these environments. Reasons for this that emerged over time included:

1. *Social context violations.* Serious problems emerged with the presumed ease of translating the "office drop-in" model of use for Cruiser videoconferencing in a corporate work environment to classrooms in schools. In the workplace, Cruiser was used to create extended workspaces, helping to foster the kind of informal interaction that people have while "cruising" the hallways of their office space (Fish *et al.*, 1993). Teachers did not like the thought that somebody could interrupt their classroom activity without warning. So Cruiser stations in classrooms were as a matter of practice not "left on" as they were in the offices of Bellcore workers.

2. *Issues of coordination with distant partners.* One problem that surfaced had to do with the tight windows of available time that any single classroom could be on-line. Whereas in the workplace environment one might reasonably expect to find someone at her desk at any time during the day, in the classroom environment particular students are only typically present for 40 minutes at a time. Although it may be possible for a scientist mentor to adjust her schedule to be available during a particular class period, the problem becomes much worse when two teachers or students at different schools wish to communicate. As school time is rigidly structured, special arrangements must be made to facilitate interaction between two parties who do not happen to have class during the same hour.

3. *Adolescents and school culture.* Teachers were worried that adolescents would use desktop videoconferencing as a medium for socializing and ignore their project work, so they were leery to incorporate videoconferencing into project activities or their requirements. This limitation in access for the students meant that it was not used in their investigative work as we had anticipated. It violated classroom communicative practices.

Furthermore, few cross-school projects developed, as it turned out that teachers did not have a strong enough motivation in favor of cross-school collaboration by students to overcome the challenges that such collaboration presented. For example, the teachers had never before needed to negotiate standards for assessing student work on projects across schools. Without common grounds for assessment, they were less likely to encourage cross-school student collaborations. Reflecting this difficulty, video conferencing was not used within cross-school collaborations. The same limitation prevented the expected use of Cruiser to extend the audience for presentations of students' projects.

4. *Lack of critical mass of users.* Another coordination challenge was presented by the fact that Cruiser did not support communication between CoVis students and relevant outside populations. Although there were some scientists using Cruiser that students could communicate with (at the Exploratorium and UIUC), the vast majority of students established connections to outside mentors and others using e-mail, which was more widely available and allowed for students and outsiders to communicate on their own, nonsynchronized schedules. Thus, Cruiser was also prevented from flourishing because of the lack of what media theorists have called "critical mass" (Markus, 1987).

The switch to CU-SeeMe helped solve some of the problems described above, but not all of them. During the 1995–96 academic year, videoconferencing using CU-SeeMe began to take root as part of regular CoVis activities. Although the cultural issues of synchronous tools were still present, the associated cost and complexity issues were greatly reduced, and the critical mass problems were minimized. We worked to ameliorate the cultural problems caused by synchronous communication through scaffolding and structuring its use in interschool curricular activities.

One example of videoconferencing activity embedded within a larger curricular context was a model summit on global climate change, held as a culminating activity for a month-long set of activities on climate change (Gordin *et al.*, 1996b).

We have also been able to use desktop videoconferencing on CoVis workstations to provide live daily "interactive weather briefings" from atmospheric researchers (faculty and students) at the UIUC Weather Room. Such briefings can illustrate, through interpretation and analysis of weather charts, satellite and radar animations, and forecast products, key concepts that will enable a student to conceptualize the structure and dynamics of the atmosphere. CoVis students participate in real-time discussions of weather processes, collaboratively interact with scientists around shared data involving atmospheric, kinematic, and dynamic processes depicted on weather charts.

A third type of activity was the "virtual field trip" between various CoVis schools and San Francisco's Exploratorium Science Museum. Using a backpack equipped with a video camera and a remote link to the Internet, museum staff in San Francisco can lead live interactive tours of various exhibits (e.g., investigating the weather section, or exhibits on erosion), as directed by students using CU-SeeMe.

A fourth type of use, unanticipated by the designers at the start of the school year, was that of CU-SeeMe to supplement parent–teacher meetings and informational sessions. This allowed staff at Northwestern to be invited "into" classrooms to answer questions about the various activities students were engaged in as part of CoVis. This combination of ways that CU-SeeMe based videoconferencing has been employed by CoVis teachers and students is an important indicator of its successfully taking root in classroom culture, especially when compared with Cruiser.

What Does This Tell Us about Design Both Generally and Specifically? Our experiences with Cruiser taught us that sociocultural barriers to use of synchronous tools in the classroom are more relevant to design of videoconferencing tools than pure technological issues, and also that cost and complexity need to be eliminated nearly completely if videoconferencing is to stand a chance at adoption. The combination of these difficulties and a lack of critical mass made it highly unlikely that Cruiser would be adopted by CoVis students and teachers. When the CoVis Project switched to CU-SeeMe, we were able to focus not on the technology but on the design of activity with embedded videoconferencing to overcome some of the social barriers.

Most importantly, the "office worker" paradigm of informal videoconferencing to the desktop does not transfer "whole cloth" to the classroom (Fishman, 1996). New design must acknowledge the different goals and social arrangements present in classroom environments than in the workplace. The social conditions for collaboration assumed to be supported by videoconferencing must be in place before the tool will be seen as serving useful functions. Furthermore, the user interfaces to videoconferencing applications for the classroom must be specially designed. Just as the Cruiser application we used was crafted with the realities of the white-collar office in mind, we have to articulate the design constraints from the classroom that should shape informal videoconferencing for teaching and learning.

What Challenges Remain? An inherent weakness of Internet-based videoconferencing systems such as CU-SeeMe is that they take up so much bandwidth that it becomes impossible to simultaneously use screen-sharing software such as Timbuktu. This poses a serious design challenge for CoVis pedagogy, which has as its goal the creation of seamless rich media environments to support collaborative learning at a distance. One possible solution to this problem may lie in the construction of Web sites where meeting participants can store the work to be discussed in advance. Another solution is being developed by Apple Computer, in its QuickTime Conferencing (QTC) software. QTC is similar to CU-SeeMe, except that it includes a shared whiteboard facility that can be used to exchange screen shots. Our experiences to date with both improving software and growing network capacity leave us optimistic about the potential for Internet-based videoconferencing as a regular feature of science education in the future.

We are finding a strong potential in uses of CoVis testbed videoconferencing in support of scheduled events that can be drawn on by teachers as dynamic classroom resources. These events—interactive weather briefings, and virtual field trips to a science museum—have allowed students and teachers to interact remotely with scientific experts in informal and formal activities. How to effectively scale network-accessible video events is an obvious challenge. Here, too, designs must be evolved for how teachers and children will make use of these events for classroom purposes. As there are no preexisting models of applications like these, a certain amount of experience-informed iteration will be necessary. Much work remains to change the ways that schools think about the division of time during the day to help make synchronous communication tools such as desktop videoconferencing more generally applicable.

An Educationally Appropriate World Wide Web Resource

Context of Problem. Exponential growth in uses of the World Wide Web (WWW) recently has also benefited education, as many K–12 schools are now regularly seeking information and resources for learning and teaching there. Schools themselves have been creating a presence on the Web, with over 5000 school-based Web sites registered worldwide as of November 1996 (Web66, 1996). But Web servers do not automatically make for educationally-useful resources. In fact, we believe that the majority of resources available on the WWW do not sufficiently meet the needs of students and teachers (Gordin et al., in press). It is interesting to note that the Web itself is one of the first major developments in computer and communication technology to make its presence felt at all levels of computer-using society, from school and work to home and community. The Web did not experience the "trickle down" that is present in so many other forms of new technology (Pea and Gomez, 1992). CoVis classrooms have been a part of the Web revolution at each step in its development, using Web resources since the early prototype versions of NCSA's Mosaic were available. The CoVis Project released its first Web server in early 1994.

Educational Needs of Learners and Teachers. Web servers that are designed with the needs of education in mind will have teaching and learning-centered design, including special purpose "front-ends" to web resources, and supporting structures that help educational users match resources to teaching–learning needs. They need to support the following resources:

1. "Just in time" activities, materials, and resources to support classroom curricula
2. Facilities to support user authoring and browsing of commentaries on resources and how they may be applied in the classroom
3. Facilities for teachers, students, and other members of a learning community to establish connections to those who share common interests
4. Powerful facilities for searching and filtering the vast array of resources generally available on the Web

An educationally focused Web server will also include suggested or tried-and-true activity structures, and sample assessment rubrics to guide teachers in the integration of activity into daily classroom life. Educational Web servers must be designed to grow organically as the needs of their user population change and mature.

What Did We Develop? In response to these challenges, we developed the CoVis Geosciences Web Server, available on-line at http://www.covis. nwu.edu/Geosciences/index.html. Introduced in the third year of the CoVis project, the design of this server is a direct result of ongoing conversations with teachers, students, geoscientists, and learning researchers about the resources they required to achieve our shared vision of science education. The Geosciences Web Server has at its core four components designed to address the needs of learners and teachers as enumerated above.

A Web section called "Activities" provides teachers with access to a growing number of curriculum units tied to the Geosciences. These units include: land use management planning, water quality, soil science, global warming, and meteorological science. Each set of materials consists of sample activities for introducing students to the topic, pointers to relevant resources, access to listserv-based discussion groups for teachers, sample assessment rubrics and links to state and national standards, and suggestions for culminating activities. It is important to stress that these materials are not the on-line equivalent of "shrink-wrapped" curricula. These materials are intended to serve as jumping-off points for teachers, and thus all require local adjustments before being useful in particular classrooms. An important role of these shared activities is that they provide the necessary shared contexts for building interconnections between classrooms. The activities posted by CoVis are intended to be seeds for a larger activities development process involving teachers. Teachers are encouraged to submit their own activities (via e-mail) and to submit their variations on the CoVis-developed activities.

The "Teachers' Lounge" Web section provides links to materials about assessment, links to materials on the philosophy of teaching project-based science and related topics, and access to several resources designed to support teachers' daily work tasks. The first of these resources is a facility that allows teachers to create and annotate personal Web pages. We call these "link lists," and they are essentially a bookmark list created by teachers for use by their students. This facility does not require teacher's to have their own server, or to know HTML. Thus, it provides an entry point for the creation of content by teachers on the Web. A second resource available here is called the "CoVis Mentor Database," and it serves the function of an "intellectual ride board" to connect scientists who want to participate in student science projects to students and teachers who seek connections and advice from beyond the classroom (O'Neill *et al.*, 1996).

A third Web section called "Resources" is a collection of on-line scientific visualization tools and links to other web servers with pertinent Geosciences information. The visualization tools (discussed in more detail in an earlier section) include a greenhouse effect visualizer, and a new Java-based interactive weather visualizer designed for CoVis by UIUC (Hall *et al.*, 1996). Although we have not yet created powerful search engines to help teachers and students make sense of the broad and emerging range of on-line resources, we have established a collection of links to resources that we know about today.

A section called "Community News" provides the important function of a "town square" for teachers and students to share information about exciting events in their classrooms and schools. This is an important community-building component of the Geosciences Web Server. It provides a window for people to learn about what their colleagues are doing and recognizes excellence.

How Did It Work in Use? Resources like Web servers are often evaluated from a technology-oriented perspective—how many "hits" do they receive? On this dimension, CoVis resources score well, receiving thousands of hits each week from uses within our CoVis community of schools, but also from schools and other places around the world. From another perspective, we know that as of November 1996, over 2000 users and organizations have created Web pages linking to the CoVis home page (and we have no way of knowing how many others bookmark the site in their Web browsers). These CoVis Web links outside the CoVis hosts imply usefulness of these resources, but the statistics do not tell how they are being used . However, we believe that a more appropriate measure of success for educational Web services is the extent to which empirical observations indicate that they serve the needs of the community for which they are designed. In this regard, our results are more mixed.

On the positive side, teachers report that these services are valuable to them in their classrooms, and when we visit classrooms, we find that the CoVis home page has been set as the default, or that the school's own home page points prominently to ours. In a sense, our user community "votes with its feet," so resources that are not perceived as valuable receive much less use than those that are seen as essential. The pages dedicated to our curriculum and activities and the scientific visualizers are the most popular, and by translation, the most useful to our teachers at the moment. Newer resources such as the Mentor Database are still being tested, and we expect their popularity to grow as well.

A downside to the usefulness of our Web services is tied not to their design but rather to limitations associated with the Web in general. In particular, the network can become sluggish at peak times of use throughout the day. This means that the most graphically intensive resources, such as the visualization tools, are often not practical for use in classes. At the heart of the issue is that Internet resource availability is not guaranteed, and frequently network problems arise at precisely the moment when teachers were hoping to use them.

What Challenges Remain? Web servers—like textbooks and other media before them—must continue to evolve so as to be useful in teaching and learning contexts. The CoVis Geosciences WWW server provides an initial step in this direction. A growing number of Web-based applications must be developed that can be appropriated by teachers, and there is much work yet to be done to design Web resources that encourage sharing and make it easy for teachers to make their own insights, experiences, and resources available to a broader community of their peers and colleagues. Ongoing issues of bandwidth limitations must also be met before educational Web services are broadly useful.

Conclusions

In the past, technologists and marketers have considered education an "application" for their existing products and technologies rather than a driver of new products and services. Our point here has been that it is more productive to think of teaching and learning as a driver of technology development. Although this perspective has been rare in the past, the development of cyberspace is an opportunity to design new underlying features and applications motivated by the needs of teaching and learning rather than seeing education as just a reason to repurpose and repackage existing applications.

We used our experience in establishing and evolving the CoVis testbed to illustrate this point (see also Gomez *et al.*, submitted). It is clear from our experience that we have had successes engaging teachers and children when we have understood their needs and created applications that fundamentally reflected teaching and learning contexts. In contrast, we have been least successful when we attempted to transfer, wholesale, an application or application-concept to schools without the necessary rethinking or fundamental reconceptualization. Our experience with Bellcore's Cruiser videoconferencing is a case in point. We believe that teachers, at least in part, shied away from videoconferencing because the form in which we presented it to them came directly from the world of white collar office work. In that world, for example, people often have private space in which to work. Personal video conferencing makes sense. In the class, private space is minimal. Therefore, informal video conferencing must be re-thought for the classroom.

On the other hand, we were more successful with the Collaboratory Notebook and scientific visualization applications because they were informed by the needs of learning communities from their earliest inception. In the case of the Collaboratory Notebook, we started with a general notion—collaborative hypermedia—and shaped the design of the application in iterations of classroom use and teachers' commentaries. In the case of the scientific visualization applications, tacit knowledge of the scientists who use such tools was made explicit and easier access to meaningful activities provided through the creation of custom visualization front-ends for specific domains of inquiry. We subsequently advanced the "fit" of scientific visualization as a tool for learning to the interests of teachers, and of learners. We did this in two ways: (1) by establishing, as teachers called for, CoVis Interschool Activities that set up a context of curriculum topics and activities to guide the uses of visualization, and (2) by reconceptualizing visualization as a medium for the classroom that emphasized students expressing their interests and beliefs through visualizations as well as interpreting the visualizations of scientists. Further, we added human geographic data sets (e.g., population) to address what we found to be preexisting student interests.

An essential requirement of our effort to create applications of cyberspace that reflect the needs of teaching and learning contexts has been evolving our view of what it means for technology to be integrally used in schools. Traditionally, the designers' goal is often thought about as one of getting technology "adopted." We see the adoption metaphor as fundamentally limited, a distorted one-way view of the route to integral technology use. The adoption perspective assumes that a technology comes packaged to a community. Shortcomings are viewed as problems with the adopters, not with the technology. A contrasting view is that technologies and other artifacts are "appropriated" by people rather than adopted (Pea, 1992). In this view technologies come to be used by individuals and communities based on a two-way process of "reciprocal evolution" (Allen, 1993). The user and designer each interpret the utility of an artifact. The artifact is then subject to cycles of iteration that reshape it based on the needs of the community of users and the designer's ongoing inventive responses to those needs (Greenbaum and Kyng, 1991; Schuler and Namioka, 1993).

The CoVis design goal is to create a tool suite shaped by a process of appropriation that includes the use of participatory design methods for refining the functionality and interface properties of new cyberspace applications coupled with intensive professional development activities for teachers as codesigners. If we are successful, teachers will innovate educational practices that are learner-centered. They will not see applications as simply being "delivered" to them for a prescribed use, but will invent uses in their local contexts that we cannot foresee.

Since we began the CoVis Project in 1992, our intentions of crafting a model collaboratory devoted to distributed science learning have led to current work with 43 schools in 11 states involving thousands of students. These schools are extraordinarily diverse in terms of socioeconomic profile, geography, and in bandwidth connectivity to the Internet. The primary challenge continues to be developing the practices and the tools of the collaboratory so that it is appropriated into teaching and learning practices, not simply "delivered" for use. In addition to the Collaboratory itself we hope to evolve design principles that will assist others in the creation of collaboratories that can be appropriated by other communities of practice.

NOTES

The CoVis Project has been funded by National Science Foundation Grants MDR-9253462 and MDR-9454729, Illinois Board of Higher Education Eisenhower grants, and our industrial partners Ameritech and Bellcore. We are also grateful for hardware and/or software contributions by Aldus, Apple Computer, Farallon Computing, Sony Corporation, Spyglass, and Sun Microsystems. An earlier version of this chapter appeared in the 1995 Proceedings of the Internet Society. For additional information, check the World Wide Web address http://www.covis.nwu.edu.

REFERENCES

C. L. Allen, "Reciprocal evolution: A design and research strategy for the development and integration of basic research, design, and studies of work practices," in D. Schuler and A. Namioka, eds., *Participatory Design*. (Hillsdale, NJ: Lawrence Erlbaum Associates, 1993), pp. 239–253.

Bellcore Information Networking Research Laboratory, "The Touring Machine System," *Communications of the Association for Computing Machinery* 36(1)(1993):68–77.

K. W. Brodie, L. A. Carpenter, R. A. Earnshaw, J. R. Gallop, R. J. Hubbold, A. M. Mumford, C. D. Osland, and P. Quarendon, *Scientific Visualization* (Berlin: Springer-Verlag, 1992).

J. S. Brown, A. Collins, and P. Duguid, "Situated cognition and the culture of learning," *Educational Researcher,* 18(1)(1989):32–42.

V. G. Cerf, A. G. W. Cameron, J. Lederberg, C. T. Russell, B. R. Schatz, P. M. B. Shames, L. S. Sproull, R. A. Weller, W. A. Wulf, *National Collaboratories: Applying Information Technology for Scientific Research* (Washington, DC: National Academy Press, 1993).

A. Collins, J. S. Brown, and S. E. Newman, "Cognitive apprenticeship: Teaching the crafts of reading, writing, and mathematics," in L. B. Resnick, ed., *Knowing, Learning, and Instruction: Essays in Honor of Robert Glaser* (Hillsdale, NJ: Erlbaum, 1989), pp. 453–494.

L. Cuban, *Teachers and Machines: The Classroom Use of Technology since 1920* (New York: Teachers College Press, 1986).

D. C. Edelson and D. K. O'Neill, "The CoVis Collaboratory Notebook: Supporting collaborative scientific inquiry," in A. Best, ed., *Proceedings of The 1994 National Educational Computing Conference* (Eugene, OR: International Society for Technology in Education, 1994), pp. 146–152.

D. C. Edelson, D. K. O'Neill, L. M. Gomez, and L. D'Amico, "A Design for Effective Support of Inquiry and Collaboration," in J. L. Schnase and E. L. Cunnius, eds., *Proceedings of CSCL '95: The First International Conference on Computer Support for Collaborative Learning.* (Hillsdale, NJ: Erlbaum, 1995a).

D. C. Edelson, R. D. Pea, and L. Gomez, "Constructivism in the collaboratory," in B. G. Wilson, ed., *Constructivist Learning Environments: Case Studies in Instructional Design* (Englewood Cliffs, NJ: Educational Technology Publications, 1995b), pp. 151–164.

R. S. Fish, R. E. Kraut, R. W. Root, and R. E. Rice, "Video as a technology for informal communication," *Communications of the Association for Computing Machinery* 36(1)(1993):48–61.

C. Fisher, D. C. Dwyer, and K. Yocam, eds., *Education and Technology: Reflections on Computing in Classrooms* (San Francisco, CA: Jossey-Bass Publishers, 1996).

B. J. Fishman, High-end high school communication: Tool use practices of students in a networked environment, Unpublished doctoral dissertation, Northwestern University, Evanston, IL (1996).

B. Fishman and L. D'Amico, "Which way will the wind blow? Networked computer tools for studying the weather," in T. Ottmann and I. Tomek, eds., *Educational Multimedia and Hypermedia, 1994*, Proceedings of Ed-Media'94 (Charlottesville, VA: AACE, 1994), pp. 209–216.

L. M. Gomez, B. Fishman, and J. Polman, "Media spaces and their application in K–12 and college learning communities," *Proceedings of the 1994 Conference on Computer–Human Interaction* (Conference Companion) (New York: ACM Press, 1994), pp. 185–186.

L. M. Gomez, B. J. Fishman, and R. D. Pea, "The CoVis Project: Building a large scale science education testbed," *Interactive Learning Environments* (in press).

D. Gordin and R. D. Pea, "Prospects for scientific visualization as an educational technology," *Journal of the Learning Sciences*, 4(3)(1995):249–279.

D. N. Gordin, J. L. Polman, and R. D. Pea, "The Climate Visualizer: Sense-making through scientific visualization," *Journal of Science Education and Technology*, 3(1994):203–226.

D. N. Gordin, D. C. Edelson, and R. D. Pea, "The Greenhouse Effect Visualizer: A tool for the science classroom," *Proceedings of the Fourth American Meteorological Society Education Symposium*, Dallas (1995).

D. N. Gordin, D. C. Edelson, and L. Gomez, "Scientific visualization as an interpretive and expressive medium," *Proceedings of the International Conference on the Learning Sciences* (Evanston, IL: Northwestern University, 1996a).

D. N. Gordin, D. C. Edelson, L. M. Gomez, M. Lento, and R. D. Pea, "Student conference on global warming: A collaborative network-supported ecologically hierarchic geosciences curriculum," *Proceedings of the Fifth American Meteorological Society Education Symposium*, Atlanta (1996b).

D. N. Gordin, D. C. Edelson, and R. D. Pea, *Supporting students' science inquiry through scientific visualization activities,* Paper presented at the American Educational Research Association, New York (1996c).

D. Gordin, L. Gomez, R. D. Pea, and B. Fishman, "Using the World Wide Web to build learning communities in K–12," *Journal of Computer-Mediated Communication* 2(3)(1996).

J. Greenbaum and M. Kyng, eds., *Design at work: Cooperative Design of Computer Systems* (Hillsdale, NJ: Lawrence Erlbaum, 1991).

S. Hall, M. Ramamurthy, R. Wilhelmson, D. Plutchak, D. Wojtowicz, and M. Sridhar, *The Weather Visualizer: A Java Tool for Interactive Learning,* Proceedings of the Education Initiatives Symposium, International Geoscience and Remote Sensing Symposium, Lincoln, NE (1996).

S. Heaviside, E. Farris, G. Malitz, and J. Carpenter, *Advanced telecommunications in U.S. Public Elementary and Secondary Schools, 1995* (Fast Response Survey System NCES 96-854) (Washington, DC: National Center for Education Statistics, U.S. Department of Education, 1996).

J. Hawkins and R. D. Pea, "Tools for bridging everyday and scientific thinking," *Journal for Research in Science Teaching*, 24(4)(1987):291–307.

J. Lave and E. Wenger, *Situated Learning: Legitimate Peripheral Participation* (New York: Cambridge University Press, 1991).

J. Lederberg and K. Uncapher, eds., *Towards a National Collaboratory: Report of an Invitational Workshop at the Rockefeller University, March 17–18, 1989* (Washington DC: National Science Foundation Directorate for Computer and Information Science, 1989).

M. C. Linn, N. B. Songer, and B. Eylon, "Shifts and convergences in science learning and instruction," in D. Berliner and R. Calfee, eds., *Handbook of Educational Psychology* (New York: Macmillan, 1995).

S. McGee, "Where is your data? A look at student projects in geoscience," *Proceedings of the Fourth Symposium on Education at the 75th Annual Meeting of the American Meteorological Society* (1995).

S. McGee, Designing curriculum based on science communities of practice, Unpublished doctoral dissertation, Northwestern University, Evanston, IL (1996).

S. McGee and R. D. Pea, "Cyclone in the classroom: Bringing the atmospheric sciences community into the high school," in *Proceedings of the Third American Meteorological Society Symposium on Education, 74th Annual Meeting of the AMS* (Nashville TN: American Meteorological Society, 1994), pp. 23–26.

McKinsey & Co., *Connecting K–12 Schools to the Information Superhighway* (Palo Alto, CA, 1995).

M. L. Markus, "Toward a 'critical mass' theory of interactive media: Universal access, interdependence, and diffusion," *Communication Research*, 14(5)(1987):491–511.

B. Means, ed., *Technology and Education Reform: The Reality behind the Promise* (San Francisco, CA: Jossey Bass, 1994).

B. Nardi, ed., *Context and Consciousness: Activity Theory and Human-Computer Interaction* (Cambridge, MA: MIT Press, 1996).

National Center for Education Statistics, *Advanced Telecommunications in U.S. Public Elementary and Secondary Schools, 1995* (Washington, DC: U.S. Department of Education, 1996). (http://www.ed.gov/Technology/starrpt.html)

National Research Council, *The National Science Education Standards* (Washington, DC: National Academy Press, 1996).

D. Norman, and S. Draper, eds., *User-Centered System Design* (Hillsdale, NJ: Lawrence Erlbaum, 1986).

D. K. O'Neill, D. C. Edelson, L. M. Gomez, and L. D'Amico, "Learning to Weave Collaborative Hypermedia into Classroom Practice," in J. L. Schnase and E. L. Cunnius, eds., *Proceedings of CSCL '95: The First International Conference on Computer Support for Collaborative Learning* (Hillsdale, NJ: Lawrence Erlbaum, 1995).

D. K. O'Neill, R. Wagner, and L. M. Gomez, "Online mentors: Experimenting in science class," *Educational Leadership*, 54(3)(1996):39–42.

S. Papert, *The Children's Machine: Rethinking School in the Age of the Computer* (New York, NY: Basic Books, 1994).

R. D. Pea, "Augmenting the discourse of learning with computer-based learning environments," in E. de Corte, M. Linn, and L. Verschaffel, eds., *Computer-Based Learning Environments and Problem-Solving* (NATO Series, subseries F: Computer and System Sciences) (Berlin: Springer-Verlag, 1992), pp. 313–343.

R. D. Pea, "The Collaborative Visualization Project," *Communications of the Association for Computing Machinery*, 36(5)(1993):60–63.

R. D. Pea, "Seeing what we build together: Distributed multimedia learning environments for transformative communications," *Journal of the Learning Sciences*, 3(3)(1994):283–298.

R. D. Pea and L. M. Gomez, "Distributed multimedia learning environments: Why and how?" *Interactive Learning Environments*, 2(2)(1992):73–109.

M. Ramamurthy, R. Wilhelmson, S. Hall, M. Sridhar, and J. Kemp, "Networked multimedia systems and collaborative visualization," in *Proceedings of the Third Symposium on Education*, 74th Annual Meeting of the American Meteorological Society (Nashville, TN: American Meteorological Society, 1994), pp. J30–J33.

M. Ramamurthy, R. Wilhelmson, S. Hall, J. Plutchak, M. Sridhar, B. Fishman, D. Gordin, R. D. Pea, and L. M. Gomez, "CoVis Geosciences Web Server: An Internet-based resource for the K–12 community," *Proceedings of the Fifth American Meteorological Society Education Symposium* (1996).

R. W. Riley, M. M. Kunin, M. S. Smith, and L. G. Roberts, *Getting America's Students Ready for the 21st Century: Meeting the Technology Literacy Challenge—A Report to the Nation on Technology and Education* (Washington, DC: U.S. Department of Education, 1996). (http://www. ed. gov/Technology/Plan/NatTechPlan/title.html)

R. Ruopp, S. Gal, B. Drayton, and M. Pfister, *LabNet: Toward a Community of Practice* (Hillsdale, NJ: Lawrence Erlbaum, 1993).

D. Schuler and A. Namioka, eds., *Participatory Design: Principles and Practices.* (Hillsdale, NJ: Lawrence Erlbaum, 1993).

N. B. Songer, "Knowledge construction through exchange and dialogue: A case study of Kids as Global Scientists," *Journal of the Learning Sciences* (1995).

K. Tobin, ed., *The Practice of Constructivism in Science Education* (Washington, DC: AAAS Press, 1993).

D. Tyack and L. Cuban, *Tinkering Toward Utopia: A Century of Public School Reform.* (Cambridge, MA: Harvard University Press, 1995).

U.S. Congress Office of Technology Assessment, *Power On! New Tools for Teaching and Learning* (Washington, DC: U.S. Office of Technology Assessment, 1987).

U.S. Congress Office of Technology Assessment, *Teachers & Technology: Making the Connection* (OTA-EHR-616) (Washington, DC: U.S. Government Printing Office, 1995).

Web66., International WWW school registry [online document] (1996). Available: http://web66.coled.umn.edu/schools/stats/stats.html.

R. Wolff, and L. Yaeger, *Visualization of Natural Phenomena* (Berlin: Springer-Verlag, 1993). Includes CD-ROM.

CHAPTER 12

Image Processing
An Enabling Technology for Science Research, Teaching, and Learning

Richard Greenberg

INTRODUCTION

Images are a powerful and efficient way to represent and transmit information. They contain vast amounts of information, as can be formally demonstrated by considering how many bits are needed to encode a single image: it can be encoded by slicing it vertically and horizontally into picture elements (pixels), with numerical values specifying brightness and color at each pixel. An image sliced into a million pixels, comparable to the number of detectors in the human retina, thus contains about 10 million bits of information. Moreover, the human brain can receive that much information in the form of an image in a fraction of a second, roughly a million times faster than written or spoken language. Images are the fastest and most efficient way to deliver information to the human brain.

It is not surprising that experienced teachers widely and frequently use pictures as a fundamental instructional medium. However, their effectiveness does not derive solely from the technical data transmission rate. Cognitive issues such as understanding, retention, and content value deserve at least as much consideration as the formal bit rate. These issues are only beginning to be explored by cognitive science. Nevertheless, common experience has led most effective communicators to recognize and use the power of images.

Until recently, the form of images allowed students and their teachers to do little more than look at them. Imagery was available only in

Richard Greenberg, University of Arizona, Tucson, AZ 87519-0092.; greenberg@lpl.arizona.edu or http://ipt.lpl.arizona.edu

analog form (e.g., inks, paints, photos, video), which is static and relatively limited in information content.

Now the recent technology of digital image processing, designed originally as a scientific research tool, allows students and their teachers to go beyond simply looking at images. Once an image is encoded numerically, common classroom computers can do the arithmetic necessary to enhance, measure, combine, analyze, display, and animate images of photographic quality. Students can extract meaning and see things that no one had seen before. This technology is being introduced to the national science teaching community by the Image Processing for Teaching (IPT) project from the University of Arizona. IPT is now widely used on available school computers across the United States.

Sources of digital imagery are easy to find today. Federal research programs publish CD-ROMs with imagery from every field of science. Students generate their own digital images by scanning printed material or digitizing video. The IPT project itself provides large image data sets keyed to its curriculum materials developments. And images are readily available via the World Wide Web on the Internet.

The Web and other user-friendly structures on the Internet have emerged during the same period as digital image processing became an accessible technology. The Internet has captured the public imagination by opening vast channels for communication and for easy exploration of data and other files on computers around the world. Educational applications seem obvious, including access to information and interaction with students, scholars, and researchers at distant locations.

However, a certain amount of skepticism will be appropriate until the true value of this technology is clear. Schools still remain largely unconnected, accessible data are randomly distributed and often invalid or spurious, the commitment of scientists to working with schoolchildren is untested, constraints on teachers (time, preparation, curriculum, and so forth) remain overwhelming, and the World Wide Web itself is groaning under its unanticipated load. A backlash by discouraged teachers is a real possibility. Nevertheless, if we assume that these difficulties will be overcome (which after all is the premise of this book), we must consider the kinds of information that are most widely accessed on the Web and how students can use it.

Images are the dominant form of data accessed on the Web. They are attractive to users and information providers know it. (Another reason they dominate is that they contain so much information, consuming a large fraction of the bit capacity of the system.)

Students and teachers who have learned to use IPT recognize the value of the digital form in which all images from the Internet arrive at their computers. They are equipped to go beyond simply viewing these images and are able to use the full power of image processing available on their classroom microcomputers.

On the other hand, for most users, the miracle of the Web is the great accessibility it gives to such a vast array of information. They are satisfied to be able to receive these images and look at them. Once an image arrives at their computer, it is simply viewed as if it were an analog (video or printed) image. Unfortunately, many information providers make images available with such limited, qualitative end usage in mind. Consequently, imagery is distributed with less than its full information content. Usually this is done to compress the data, speeding transmission and facilitating storage. Video provides the standard of quality for much of the image traffic on the Web.

Another reason that images are purposely provided in degraded form is more insidious. Original scientific data are very valuable to the researchers who obtain the data. They often deliberately degrade images for distribution on the Web for public information or educational consumption. For example, the chairman of the committee overseeing data processing for a major federal research program recently wrote that the challenge in designing distribution of data over the Web for educational use is to make the images attractive, while ensuring that they are in a format that is scientifically useless.

Despite such attitudes, valid digital imagery is widely available to students and teachers. IPT provides them with a state-of-the-art scientific research tool for investigation of real, rich, complex data sets. Students are equipped to learn through personal exploration and discovery.

WHAT IS IMAGE PROCESSING?

Until the early 1980s, through much of NASA's Voyager project, imaging data radioed back to Earth by U.S. interplanetary spacecraft were converted into photographs at a central computer facility at the Jet Propulsion Laboratory. The data were taken with an early digital camera, which used a detector to measure the brightness at every point in its field of view. The numerical values returned to Earth were converted to an analog product. This procedure was expensive and slow. Moreover, the photos delivered to the scientific community contained only a fraction of the information in the original data, often hiding important features. For example, images of Jupiter's moon Io taken by the Voyager spacecraft in 1979 showed a puzzlingly weird surface, which the science team could not explain. Only after the same digital data were reprocessed in the image processing laboratory to enhance the background sky for the optical navigation team, an optical navigation engineer discovered the volcanic plumes that explained everything. These features were always in the digital image data, but could only be seen when processed serendipitously.

During the 1980s, the technology of digital image processing moved from such cumbersome origins in a few specialized computer facilities to widespread applications, even using common microcomputers. By the end of that decade, the Io-volcanism discovery could be reproduced in seconds using a Macintosh computer. This progress was driven to a great extent by the needs of planetary exploration, where we want to make best use of very rare and expensive spacecraft data, and of biomedicine, where we want to extract information while minimizing invasion into living bodies.

Image processing is possible on a computer because it deals with digital data: The images are stored as arrays of numbers, like the data radioed back from spacecraft. Therefore image processing allows pictures of photographic quality to be manipulated through a computer. Processing can include measurement, contrast enhancement, false coloring, animation, filtering, and unlimited quantitative analysis. This technology offers the ability to manipulate images to bring out features and properties that had previously been difficult or impossible to perceive. It facilitates the manipulation of data in ways that revolutionize how information is perceived, analyzed, communicated, and stored. Image processing is now an important tool in many major areas of scientific study.

Currently, most images used in education are still in analog form, like the photographs that frustrated planetary scientists in the early years of the Voyager mission. Even with modern multimedia systems, computers are essentially used as catalog and retrieval devices for analog images. Multimedia presents images to users, but generally does not allow them to modify and process the images.

With digital image processing, users become empowered and engaged in ways that are not possible with other image media, just as scientists did beginning in the 1980s. The technology has advanced to the point where sophisticated state-of-the-art public domain software runs on microcomputers which are now becoming widely available in schools.

PREMISE OF THE IPT PROJECT

In 1989 at the University of Arizona, we realized that image processing might be relevant to education for a variety of reasons.

Independent of content, image processing itself combines high technology, information theory, and cognitive issues. Accordingly, it seemed a natural technology for teaching and learning. Just as word processing and other computer applications were finding increased roles in the classroom, image processing seemed a logical (and probably inevitable) step. Moreover, the content areas to which image processing is being applied in the worlds of science and technology are intrinsically interesting themselves and span the subjects of school curricula and student interest.

Image processing also fits naturally into the educational idea of "constructivism." There seem to be numerous definitions of constructivism, as this expression is bandied about, but one definition (Yager, 1991) makes a great deal of sense to a research scientist like myself, based on the following considerations.

It seems reasonable to believe that children naturally learn by a process similar to scientific research. Starting as infants they explore the world, having experiences and making observations. They construct sets of rules that impose some pattern to their experiences and allow them to make useful predictions about what may happen under various circumstances. When additional experiences or observations fail to fit their schema, they are forced to revise or reconstruct them.

Scientists follow the same procedure, constructing theories based on experiments and observations, testing the theories with further experiment and observation. The real fun is when the results do not fit, and the theory must be modified or completely reconstructed. A classic example was the fall of Newtonian mechanics and the construction of relativity.

It follows that science is what children do naturally. Students entering a science classroom have been doing science all of their lives, so they bring with them their own scientific theories. These theories are sometimes called "misconceptions," but in fact they are no more misconceptions than Newtonian mechanics was.

A common, perhaps prevalent, way of teaching science involves requiring students to memorize facts and learn other peoples' theories, demonstrating proficiency by performing practiced tasks and responses. That teaching method seems to follow the "behaviorist" tradition of Pavlov and Skinner: Ring a bell and dogs salivate; ask a question and kids give the answer. There are a number of problems with that approach: Even students who perform very well may never change their underlying conceptions about how the world works; potential scientists are most likely to be repelled by the process and perform poorly; and all students miss the opportunity to practice the processes of science.

In contrast, constructivist education (in the definition most compelling to a scientist) guides students to do experiments and make observations that may clash with their preconceptions, forcing them to change their underlying worldview, in much the same way as research scientists update and revise theories. This process brings on a fundamental change in students' knowledge of their world, going far beyond what can be measured on a conventional test.

Image processing's great potential as an educational tool comes from the fact that it was invented to facilitate the processes of exploration and discovery in research. In the classroom, image processing offers students meaningful ways to investigate imaging data that are too extensive to have been fully explored by the scientific community. It offers all students the power to see those data (or data they generate themselves) in novel and individualistic ways. The door is open to original scientific discovery. Therefore, IPT is an intrinsically "constructivist" medium when used by schoolchildren. Effective use of IPT in a classroom requires setting aside common "behaviorist" ways of using technology for teaching.

Another hypothesis early in the IPT project was that, for visual learners (which may include everyone for the reasons discussed in the introduction), manipulation of images may provide a more attractive entree into science and mathematics than traditional language-based ways of introducing and teaching these subjects.

Traditional language-based teaching methods in science and mathematics have generally failed to inspire students from diverse linguistic and cultural backgrounds. We suspected that learning through image manipulation might be a particularly significant opportunity for these students. The subjects of imaging data can provide an inviting familiarity (e.g., images from the students' experience or environment), as well as access to the frontiers of human exploration and high technology. Learning through image manipulation complements and enhances language-based instruction for all students.

To test our hypotheses, we developed an initial project in which selected teachers would learn image processing during extensive summer workshops and develop ways that their students could use it in their classrooms. Each teacher would be sent back to school with a computer to try the activities developed by all participants. What made this experiment possible was that in 1989 image processing was just becoming possible on microcomputers, and most significantly the software *NIH Image* had become available.

Why was this system so appropriate? *NIH Image* is a real research tool, designed for the biomedical research community (and now widely used in all fields of science), meeting our goal of giving children a real tool for research. Moreover, it is a tool students can easily use, with a friendly interface. In contrast, most other image processors did and still do involve parametric line commands and other coded input, not much of an improvement over the kinds of code that traditionally turns kids away from science and mathematics. Finally, *NIH Image* is in the public domain, developed and maintained by the National Institutes of Health.

Thus a state-of-the-art technology was readily available for any school that chose to use it. In what follows I describe how we addressed in sequence two major issues: finding out whether image processing really has a useful role in the classroom and implementing IPT as a significant, widely used program.

THE IPT EXPERIMENT

To explore the potential of digital image processing as a medium for learning, the IPT project first educated a group of exemplary teachers in the technique and technology of image processing, and in the scientific content of a wide variety of data sets, during a series of summer workshops. More than 80 teachers participated in the 4-week workshops of the initial experimental and school-based trial phase of the program. They served as full-fledged members of the project team, as our experts on practical education issues and on the implementation of image processing in specific school contexts.

These team members developed activities for children in a wide range of grades, subject areas, locations, and kinds of schools. Most significantly, they tested those curriculum units in their schools. They reported and compared results, and participated in the evaluation process.

The teachers, and their students in school, worked with Macintosh computers with CD-ROM reader, the software *NIH Image*, extensive data sets from a wide variety of disciplines, and activity units designed and tested by our expert teachers. Even early in the project, much of this material was distributed on CD-ROMs produced by the IPT project and from other sources, which was simply cheaper than the cost of an equivalent number of floppy diskettes. Although CD-ROM technology has achieved a mystique of its own, it is worth retaining the perspective that (apart from mechanical details) a CD is just like a fast, superhigh-capacity, read-only floppy disk.

We began this project as an experiment. It seemed plausible that image processing would serve as an effective and exciting way to attract children to science and to offer unlimited possibilities for exploration and discovery. On the other hand, we recognized that not all good ideas come to fruition in practice.

Thus, it was gratifying that evidence pointed to remarkable effectiveness of image processing in education. Exciting things happen with IPT in the classroom. Image manipulation is attractive and fun, but it inevitably leads to exploration of the scientific content, and appreciation of the mathematics that is built into the process on many levels. IPT nurtures the investigative processes of science in children. We find great success and interest by children who might not have been expected to do well in conventionally taught science classes.

IMPACT OF IPT IN SCHOOLS

The extensive follow-up and evaluation process built into the IPT project has dramatically demonstrated a number of results, some of which were unexpected (Greenberg *et al.*, 1993).

1. *Underrepresented groups.* The project is proving to be an exceptional way to reach previously under-represented groups of students, a fundamental goal of the IPT project. We are having extraordinary success with minority groups, females, children with limited English proficiency, so-called "learning disabled" children, and others. With image processing, traditional obstacles to doing mathematics and science are largely eliminated, so the playing field becomes level for participation by all. Students from underrepresented groups often become leaders in IPT, opening the door to broader successes. At the same time, for gifted and talented students from all groups, IPT offers challenging

and unlimited possibilities for advancement and enrichment with this professional scientific research tool. The IPT project has stressed the inclusion of underrepresented populations, with a strong presence ranging from rural Native American schools (e.g., on the Navajo and Tohono O'odham reservations; see Greenberg and Franklin, 1992) to inner city schools with large percentages of minority students (e.g., in Baltimore, St. Louis, Chicago, and Los Angeles). Nationwide, over 40% of the students of IPT teachers are members of underrepresented minority groups.

2. *Mathematics.* Although our original expectation was that IPT would have an impact on teaching science, we have been impressed at the effect it has on motivating and enriching study in mathematics. Our participating teachers realized this immediately. Independent of the content of an image (many of our teachers introduce image processing with images of cars, rockets, or the children themselves), the manipulation is rich in mathematics. Just by playing, students become familiar with coordinate systems, numerical arrays, scale, histograms, look-up tables, and so on. Some teachers use the gray scale map to teach concepts of slope (contrast) and intercept (brightness). More advanced students delve into the technical details of the software and quantitative analysis to a level much deeper than anticipated.

3. *Students' outlook on life.* A repeating pattern at a wide variety of our participating schools involves numerous "at-risk" students whose negative views of themselves and of the world have been radically changed by (incredibly) image processing under the guidance of our expert teachers. Students viewed as being on paths leading to academic failure (even headed toward jail) are now highly motivated to do extra work in image processing, to take leadership roles in training other students, and to become serious students. The appeal of images, the empowerment of image processing, and the opportunity to become expert build self-esteem to a degree we had not anticipated.

4. *Effects on teachers.* This project was begun to explore the potential role of this new technology in education. The teacher education part was regarded as a means to that end. However, from our various feedback and evaluation efforts, we have found that teachers' perceptions of subject matter, pedagogy, and their own professional role have been profoundly affected in some fairly consistent and important ways. In the IPT project many teachers are full members of the project team, using their expertise to continually find new ways of applying this technology to the context of their classrooms. The project is thus modeled on the "creative anarchy" of the research world, rather than the authoritarian, hierarchical structure that often characterizes the teachers' workplace. In this way we model a "constructivist" approach

to learning. Participating teachers have become acculturated to the social and intellectual processes of scientific inquiry. As a result:

a. Many appreciate the close relation between their work and the work of researchers.
b. They have incorporated this constructivist approach into their teaching.
c. Many become activists for such change in their schools.
d. Professional motivation is revitalized both among the veteran and the more junior teacher-participants.
e. Of particular interest in the context of this book, the IPT project has stimulated teachers to develop cooperative image-based projects among their geographically distributed classes and with members of the scientific research community.

5. *Middle-school success.* In its experimental spirit, our project has included teachers from the upper elementary grade levels through high school. Given the objective of introducing high technology, we anticipated that IPT would be most effective at the upper grade levels of high school. In fact, IPT is proving to have the greatest impact at the middle-school levels for two reasons: First, science curriculum is not rigidly defined for those grades, so teachers have flexibility to incorporate innovations. Second, the students are much more open to the spirit of exploration and discovery. In contrast, high-school physics and chemistry are bound to a traditional curriculum, with students who are already locked into the grade- and test-oriented system, i.e., the winners in the behaviorist game. (This result has a disturbing corollary: Many schools may have discouraged the potentially most creative scientists before they reach the upper grade levels.) Our teachers have successfully introduced IPT at all grade levels and subjects, but their greatest success at integration into the curriculum and inspiring students has been in the middle school grade range.

6. *Scientific and technical contributions by students and teachers.* IPT teachers are our experts on practical education issues, but they have gone much further. Equipped with this state-of-the-art technology, IPT teachers and their students have had an impact on the research community. They have, for example, introduced image processing to tree-ring research, influenced the continuing development of software at NIH, provided a standard system adopted by NASA Planetary Geology and Geophysics, developed new techniques for three-dimensional imaging, and contributed original research papers at NASA's Lunar and Planetary Science Conference. IPT students and teachers have been responsible for generating innovative data sets far beyond those provided by the IPT project. The significant communication

among teachers, students, and professional researchers has enriched both groups. This interchange is a strong and unique feature of the IPT project; it is an example of successful educational use of the Internet and has major implications for modeling cooperation among students and professional scientists.

7. *Most effective teachers.* As part of our experimental approach, our original team of teacher participants was selected to include a variety of different subject area specialties, grade levels, and background strengths and weaknesses. Initially, we expected that the most effective participants would be those with the strongest scientific and technical backgrounds. In fact, we find that other characteristics are at least as important for effective integration of IPT into schoolwork and curriculum. In this regard, our most successful teachers are those most pedagogically enlightened and flexible, with the energy, enthusiasm and administrative acumen to introduce this fundamentally constructivist medium into their classrooms.

8. *Availability of technology.* When we began this project, the standard hardware required for each of our teachers cost about $6000. With NSF funding, a substantial Apple donation, and cost-sharing by participating schools, this price was acceptable for our initial test project, but it was not consistent with widespread dissemination. The price of the hardware has dropped drastically, and many school districts are now buying large quantities of these machines. We have documented numerous cases of districts (not associated with IPT) that have purchased these machines, but are not using them effectively or at all! In all of these cases the need is for meaningful applications and teacher education, and IPT is becoming widely recognized as one of the most effective and meaningful uses of this technology.

9. *National recognition.* Because the power of IPT is readily recognized by teachers who see it demonstrated, word of the effectiveness of this project has spread widely, both through our presentations at hundreds of national and regional conferences and through extensive media coverage. Senator Barbara Mikulski visited with some of our participating students, and specifically praised IPT during Appropriations Committee hearings on the NSF budget. Independently, Ernest Boyer of the Carnegie Foundation has written of our project and recommended to NASA that ways be found to disseminate our materials and educate teachers in their use. The Far West Regional Laboratory has selected the IPT Project as a "Promising Practice." The NSF report "Signal Processing for the NII" (Gray *et al.*, 1995) identifies IPT as a key example of accomplishments and potential for educational use of the National Information Infrastructure. Most importantly, teachers have been spreading word of IPT among their colleagues.

DISSEMINATION

Just as the experimental phase of the project began to demonstrate the effectiveness and relevance of image processing in education, the availability of the appropriate hardware in schools became widespread. Therefore, several years ago we began the current dissemination phase of IPT. The activities of this phase are guided by the results of our ongoing evaluation of all components of the project, which confirm the general results of the experimental phase and continue to inform workshop improvement and materials development efforts through classroom testing.

It is critical to recognize that simply making data accessible on the Internet or through other media is not enough to ensure that they will be used in education, even as classrooms begin to be connected. These data will only be broadly used by the K–12 educational community if the following support is provided: (1) widespread teacher education, (2) curriculum-based materials development, and (3) substantial follow-up support.

IPT has already developed the infrastructure for broad dissemination of those essential components. Thousands of teachers have participated in multiday workshops in IPT at sites across the country; curriculum materials are available in CD-ROM form and allow for expansion by those students who have access to imaging data on the Internet; Extensive follow-up support and outreach are readily available through telephone and e-mail hotlines, the annual IPT conference, newsletters, and school visits. Thus, the IPT project serves as a full-scale model of successful implementation of massive widespread usage of the kinds of information that are becoming accessible.

Two key features of IPT are (1) the role of continual formative evaluation to inform ongoing refinements and improvements and (2) the central role of expert teachers in carrying out all of the proposed activities. We also use the expertise of the existing IPT infrastructure, continuing our strong links with the research communities that generate, use, and archive imaging data, increasingly in accessible digital libraries.

All IPT development and dissemination continues to be naturally and explicitly aligned with emerging national standards in science, mathematics, and technology education. Our materials are keyed to NCTM Standards and are aligned with the developing National Science Education Standards and with the technological education goals articulated in the SCANS reports and other national policy guidelines. All of these connections are natural, because the underlying motivation and philosophy of IPT is similar to those of national reform efforts. For example, IPT Director Greenberg was an author of the Arizona science framework document, an earlier statewide effort that foreshadowed the national standards. Moreover, our development is guided by national policy on educational applications of technology and the NII, including for example NSF's "Principles and Goals of Technology in Education" (Sabelli *et al.*, 1994) and NCC-TET's "NII: Requirements for Education."

The components of the IPT dissemination, teacher education, follow-up and outreach, and materials development and distribution, are described below.

Teacher Education

The extensive 4-week workshop format of the IPT experimental phase was not practical for the widespread teacher education needed for IPT to meet its full potential. Based on our experience we developed in-service workshops of 3 to 5 days' duration. This shorter, transportable version is in line with the amount of in-service training included in well-planned school-district technology programs, while providing the necessary theoretical background, technical training, and scientific context.

IPT workshops bring more teachers the skills they need to bring image processing to their classrooms. The OTA report on Teachers and Technology (OTA, 1995) reinforces our finding that teacher education is an essential (and frequently overlooked) key to technology implementation. IPT workshops model the constructivist processes of exploration and discovery that have been so successful in our past workshops and in the classrooms where IPT has been implemented. Most IPT instructors have considerable experience working as teachers with IPT in their own classrooms; In every case, at least one instructor is an experienced IPT teacher.

IPT workshops are designed for groups of up to 20 teachers meeting together for 3 to 5 days. The curriculum provides (1) theoretical background on image processing, (2) skills in image processing by doing many of the IPT activities designed for their students, (3) familiarity with the available set of curriculum materials and ways to customize them, (4) techniques for generation of images by the teachers and students themselves (e.g., scanning, digitized video), and (5) ways to access digital image data sets and supporting materials from elsewhere. We now emphasize even more strongly the use of the Internet for access to digital libraries, for cooperative projects among schools, for connectivity to research resources, and for exchange of student-generated data.

Fully 85% of IPT workshops are funded directly by school districts (from local in-service education funds or special grants) or by other science-education programs, providing strong leverage to the seed support for workshops provided by NSF. IPT teacher education has proven itself in the marketplace of educators' choice. Over 2500 teachers nationwide have already participated in our multiday workshops. In a recent summer, 45 workshops were held nationwide for approximately 800 participants. Using grant funds, IPT has targeted schools that represent diverse populations of teachers and learners, including groups that have been traditionally underrepresented in mathematics, science, and technology. Schools serving these groups are surprisingly often rich in hardware, but weak in applications and teacher preparedness. These schools serve as valuable test sites for our materials development efforts.

Current distance-learning and conferencing technologies are not yet adequate to replace personal contact given the intensity of IPT workshops. However, we anticipate that appropriate technologies will emerge that will allow us to conduct aspects of our teacher education with geographically distributed IPT instructors and workshop participants.

Through the IPT workshop-leader training program, some teachers develop the skills necessary to become instructors themselves. Contrary to the belief of some policymakers, teachers who have taken technology workshops are not necessarily prepared to transmit the workshop content to other teachers without such additional support and training. Therefore, we take great care in preparing those who will go on to instruct other teachers. These certified trainers can then run IPT workshops in their own districts, often coordinated with programs of systemic change in science education.

Follow-up and Outreach

A strong program of follow-up support is perhaps the most critical, but most widely neglected piece of school technology implementation. IPT provides support for teachers addressing technical, pedagogical, and scientific issues. Support is available to IPT teachers and students via the Internet, including personal e-mail hot-line access, and our World Wide Web site (http://ipt.lpl.arizona.edu). Toll-free phone support is also available. We also have an annual national conference for IPT teachers and other interested educators. Our newsletter "IPT Bits" also provides a forum for technical information and updates and for exchange of experiences with school implementation.

In addition, we have a dynamic program of outreach and publicity to inform the teaching community about the power of this new medium for learning. Demonstrations at national and regional conferences inform the teaching community about this opportunity. Many of these presentations are made by teachers who use IPT in their own classrooms. Over 120 presentations have been made at conferences nationwide; we have demonstrated IPT to over 12,000 mathematics and science teachers. We publish articles in scholarly and professional literature both to document our progress and to inform decision-makers about these opportunities. We also successfully encourage publicity for the project in the media. We have developed and maintain a large database of contacts, and widely distribute literature and other materials about IPT.

Materials Development

As part of the experimental program, our team of scientists and teachers has developed extensive activity units in IPT. These activities are designed to lead teachers and students into active inquiry and discovery, taking advantage of the power of technology that allows access to vast data sets that have only begun to be explored by the scientific community. Now we are continuing the development with an emphasis on activities that will be recognizable by teachers as fitting in with their familiar educational objectives while still allowing the exploration and personal discovery that have characterized the best IPT activities.

Materials developed so far include the workshop CD-ROM (now in version "*IPT 2.1*") containing activities in a wide range of subjects that allowed teachers to learn image processing doing the same kinds of activities that their students might do in their classes. This CD-ROM is accompanied by a manual of printed activities and reference materials, the *IPT Sourcebook*. IPT CD-ROMs all include extensive curriculum materials, associated image data sets, and libraries of imagery to explore.

The multimedia instructional package "*Hands-on Image Processing*" (HIP) , developed with a grant from the Annenberg Foundation/CPB Math and Science Project, consists of a hypermedia CD-ROM that leads a novice teacher through the fundamentals of image processing and the uses of IPT materials in the classroom. An accompanying video demonstrates how successful IPT teachers integrate this technology into their classes.

A set of subject-specific CD-ROM materials packages is under development. The first of these subject-specific packages, *HIP Physics*, is a "stand-alone" product that can be used in classrooms of teachers who might not have participated in a workshop. The impetus for this experiment was a group of IPT physics teachers, who ultimately helped develop our "*HIP Physics*" CD-ROM. This CD-ROM is proving to be very popular among innovative physics teachers at the high-school and college levels.

The recently released HIP Biology (in two volumes) has received several awards and is rapidly being adopted by biology teachers across the country.

Another recent CD-ROM product of our Teacher Enhancement dissemination phase is our "*IPT Preview*" CD which, in addition to our demo diskettes, is used to introduce potential new participants to the project. This hypermedia CD-ROM describes image processing and various aspects of the IPT project, including sample activities and images.

Our continuing curriculum materials development is guided by a set of principles, based both on the original premises of the IPT project and on what we have learned during the past 5 years.

The best of our past materials have, and all of our future materials should have, the following characteristics:

1. The materials must be curriculum-based. The use of image processing and original scientific data allows a great leap ahead of many classroom experiences. It offers the potential for activities on the forefront of research. Teachers who develop such applications, and their stu-

dents, have had extremely exciting experiences. However, we have found that for widespread use of IPT, materials must be recognizable by teachers as supporting their curricular objectives and fitting realistic scheduling constraints, while at the same time allowing open-ended exploration, discovery, and analysis. IPT activities must be innovative, but must be usable in the context of real schools and viable curriculum implementation. Each set of subject-specific activities must be designed to span a broad range of content areas within that subject, so as to provide comprehensive curriculum support and innovation.

2. The materials must provide image data sets needed for the curriculum activities, but must strongly encourage use of student-generated data. Such data may come from in-school laboratory experiences, or from spatially distributed measurement projects. Consider the following examples:

   a. IPT mitosis and cell structure activities include sets of appropriate images, but effective implementation can have students obtain their own images on "wet-lab" days: They use video cameras (now often available in school labs) to record microscopic images, then digitize the images for subsequent analysis with image processing. Students without access to such technology can use the high-quality images from the IPT CD-ROMs.

   b. The IPT motion analysis macros used in HIP Physics can be applied to the image sequences on the HIP Physics CD (such as the free-throw sequence in the *"Basketball Ballistics"*). However, students can acquire their own data sets by videotaping sports events or experiments in the gym, and then use image processing with the IPT motion macros to do the analysis (e.g., determine the acceleration caused by gravity).

   c. Other activities are enhanced and expanded from Internet exploration and research. Several examples are discussed in the section on IPT and the Internet, below.

3. The activities must use the unique capabilities of image processing. Most IPT activities depend on image enhancement and analysis that would be difficult or impossible with other image media. Even when our activities are occasionally similar to paper activities (e.g., the IPT *"Karyotyping"* activity), the medium of IPT is especially attractive to students because of the power to control the process, the speed of completion, and the connection with technical career opportunities.

4. All IPT activities are intrinsically interdisciplinary. Many educators are still bound to a tradition of keeping the subjects of mathematics, science, and technology separate from one another. In the IPT project, even when materials are developed for a specific curriculum context,

it is impossible to separate out these subjects. Such integration is a key feature of curriculum reform as articulated in emerging national standards for education in all of those subject areas.

5. IPT activities continue to support emerging national standards in science, mathematics, and technology education. Our current activities are keyed to existing standards (e.g., NCTM for our mathematics activities). Science activities will be keyed to the National Academy's Science Education Standards now that they are finalized and approved. These alignments are critical in helping teachers recognize how IPT materials naturally support their efforts to incorporate the new standards, and more fundamentally the spirit of these reforms.

6. IPT activities build on image processing as a state-of-the-art research tool to provide experiences that are constructivist in nature: They allow students to explore the complexities of the messy real world as recorded in image data, and develop their own intellectual constructs. For example, one IPT mitosis activity uses a time sequence of digital microscopic images of chromosomes separating within the nucleus. Through image processing, students construct their own schematic diagram of the process. The constructivist IPT approach contrasts with the traditional teaching of mitosis, in which students memorize a set of line drawings labeled in Greek and Latin.

7. All IPT curriculum activities must minimize cookbook instructions and open the door to exploration and discovery. As a research tool, image processing readily provides activities that encourage critical thinking, problem solving, and provide open-ended opportunities for quantitative analysis. In developing IPT activities, there is always a narrow path to be followed in providing enough guidance to direct students into avenues of personal exploration and discovery, while avoiding being overly prescriptive in instructions. Similarly, while we emphasize visual learning, our materials should help with literacy and vocabulary building. Learning to achieve the appropriate balances has not been trivial; our experience in this regard is a valuable component of what we bring to this project. The best activities provide structure to guide students, but lead to open-ended learning.

8. IPT curriculum development assumes an active role for teachers working with their students. We reject the model of drill-and-kill exercises done independently by students. IPT activities are intended to be done by small groups of students working together to explore and solve problems, and the guidance of teachers in the discovery process is crucial. Collaboration among students promotes and enhances higher-order thinking and problem-solving skills.

9. IPT materials are relevant to the real world. Many are based on current issues and recent data from scientific research. Others represent the

world of career opportunities and practical applications, where image processing plays an increasingly widespread role. Still others will relate to the daily lives of students, including sports, medical, and daily-life imagery, for example. IPT teachers who work with diverse populations have emphasized the value of incorporating culturally meaningful imagery to excite and motivate all students. Our materials testing process should ensure not only scientific and technical accuracy but educational and cultural appropriateness as well.

10. IPT materials are complementary to (but distinct from) simulation software that is increasingly used in education. It is important to clarify the distinction. IPT provides a tool for analyzing the messy, real world in all of its complexity, allowing a scientist/student to recognize and construct and connect the underlying principles and laws that govern nature. In contrast, simulation software allows a student to model an artificial world, assuming that the software incorporates the correct laws of motion or behavior. For example, Dr. Keith Mitchell of Apple Computer contrasts IPT's *HIP Physics* with the simulation package Interactive Physics: "Basketball Ballistics" (from *HIP Physics*) allows a student to extract the laws of motion from actual behavior of a ball; Interactive Physics allows a student to simulate the motion of a ball assuming that the software knows the laws of motion. Mitchell has shown how these two approaches are complementary ways for a student to learn. IPT materials should encourage integrated use of these complementary learning tools.

11. IPT materials must include built-in assessment of students' learning and achievement. Image processing naturally leads to creation of student-created products that allow teachers to track the process of exploration and discovery for an authentic assessment. Materials can be structured so as to guide students into activities and projects that yield such products, including written reports, completed projects, quantitative solutions, and digital or hard-copy image products. IPT teachers have already developed a wide range of personal approaches to assessment of critical thinking and problem-solving processes that are central to most image processing. We are building on those experiences to develop approaches that will be built in to our new materials.

IPT AND THE INTERNET

When the IPT project began, microcomputers for image processing were still prohibitively expensive. By the time IPT entered its dissemination phase, appropriate computers were becoming widely available. Similarly, in the early years of IPT, access to digital scientific imaging data was difficult for teachers; Internet access was and remains difficult for many, even for the several dozen for whom we provided modems and access accounts. We generally make imaging data available on the IPT CD-ROMs, along with the corresponding curriculum materials and software. We also have distributed image-data CD-ROMs as they became available from various research agencies, including for example NASA planetary image data.

However, as some teachers and their students have found ways to access the Internet, they have been able to add an extra dimension of exploration to their students' IPT activities:

1. Several widely used IPT curriculum units involve analysis of GOES weather satellite imagery of hurricane systems and diurnal surface temperature variations (e.g., the "*Storm that ate South Carolina*"). IPT CD-ROMs contain time-sequence data sets for these activities as well as supplemental image data for further exploration. Now consider the extension of this activity that became possible when an IPT middle-school teacher on the Tohono O'odham reservation gained Internet access: His students were able to obtain same-hour satellite data for correlation with local weather conditions. By using the Internet, the IPT activity was extended to a term-long project in weather prediction and satellite data interpretation.

2. High-school students in Arizona and Washington state have worked together via the Internet on collaborative projects analyzing and interpreting radar images of the surface of Venus from the Magellan spacecraft. The 200,000 image data set is largely unstudied because of the limited data-analysis resources of NASA. The students' results have been presented annually at NASA's Lunar and Planetary Science Conference, and are generally regarded among scientists as the definitive work on characterizing the distribution of small volcanic cones on Venus (e.g., Gerlach *et al.*, 1995).

A number of widely used IPT curricular activities are based on exploration of data primarily accessible on CD-ROMs, but the Internet allows vast expansion of the imagery available. Moreover, the best implementation of IPT activities has often been when students have created images themselves. With the Internet students exchange and compare these student-generated imaging data sets for IPT analysis. Also, new versions of images, based on students' own processing can be exchanged and compared among schools at various geographical locations. This access can provide the basis for stimulating materials across the curriculum. Thus, the Internet can enhance the open-ended exploration that is the hallmark of IPT materials.

Two examples of how IPT activities were enriched when the Internet became available in schools were given above. Other examples of existing IPT activities that have natural expansions via Internet access are:

a. *"Eyes in the Sky"* is based on multispectral Earth remote sensing data and involves schoolchildren doing image arithmetic to construct thematic and environmental maps. They can construct maps of the extent of vegetation on regional or continental scales. This activity will be opened up to the full set of remote sensing and GIS data in digital libraries of USGS, NOAA, NASA, and others, so that students can explore time variability, selected regions of interest, and particular environmental or resource issues.

b. *"Eruption Plumes of Io"* allows students to measure planetary imagery to discover that ballistic trajectories are governed by the same physical laws in space as in familiar settings. Hundreds of thousands of additional planetary images (original data largely unexplored) are accessible on the Internet via NASA's Planetary Data System for limitless exploration and discovery.

c. *"Makes my Head Spin"* uses stacked images of MRI slices (from an IPT CD-ROM) to provide three-dimensional insight into the structure of the human head; Internet access would expand the data set 1000-fold or more. With *"Enlarged Heart,"* students examine a few pathological cases on a macro and cellular scale; imagine the possibilities when students can explore numerous cases and diverse pathologies. These and many other IPT activities on anatomy, physiology, cell biology, genetics, among others, will be opened up to unlimited exploration through access to digital libraries of the National Library of Medicine and dozens of other biomedical institutions and agencies.

d. In *"Quakin' All Over"* students use image processing to plot global earthquake location data. The activity already includes an option that guides students by the Internet to data sources at the USGS so that their work can apply to the most recent seismic events, lending a sense of immediacy and excitement to the project.

CONCLUSION

The images used in IPT are not just pretty pictures. They are original data from recent research in a wide variety of disciplines and practical applications. The data are in digital form, not the analog form generally accessed through educational multimedia systems. IPT teachers and their students thus manipulate the original data bits as they do their image processing. The vast amount of data creates the continuing possibility of real scientific discovery. The process provides a learning experience based on a scientific research model, rather than the traditional school format.

The communications revolution represented by the Internet and other technological advances can provide students with access to phenomenal amounts of data, much of it in the form of imagery. It also, in principle, can allow students at diverse locations to work on cooperative projects of data acquisiton and analysis, and to include professional scientists in their collaboration. However, the technologies alone are not sufficient to ensure that this promise is fulfilled. Meaningful ways to use these capabilities must be found and made available and usable in the real world of schools.

The history of the IPT project shows how a new technology can be exploited and disseminated as a meaningful tool for learning. Although the technical aspects are challenging, equally (if not more) challenging is developing and integrating the programmatic components needed to deliver the awareness, the skills, the support, and the materials required by teachers and students in the real world of schools. IPT has succeeded thus far by concentrating its limited resources on the educational applications, not squandering them on technological development. The key has been a central and dominant role for experienced teachers in every aspect of the IPT project.

REFERENCES

R. M. Gray *et al.*, Signal processing for the NII, NSF Report 95-10 (1995).

K. A. Gerlach *et al.*, "Correlation of small volcanic edifices with geologic features on the surface of Venus," *Lunar and Planetary Science* 26(1995)495.

R. Greenberg and K. Franklin, "Image processing for teaching in a Navajo school setting," *Journal of Navajo Education* 9(2)(1992):23–29.

R. Greenberg *et al.*, "Image processing for teaching," *Journal of Science Education and Technology* 2(3)(1993):469–480.

National Coord. Committee on Technology in Education and Training, (Alexandria, VA: NCC-TET, 1994).

Office of Technology Assessment, *Teachers and Technology: Making the Connection* (Washington, DC: U.S. Government Printing Office, 1995).

N. Sabelli *et al.*, Principals and Goals of Technology in Education, Report of a workshop held October 1993 (NSF, 1994).

R. E. Yager, "The constructivist learning model," *The Science Teacher* 58(6)(1991):52–57.

APPENDIX

Student–Scientist Partnerships

Programs are listed in alphabetical order. When programs are known both by a name and an acronym, the most commonly used designation is given first. This list of programs is not meant to be exhaustive; it serves as a sample of the wide variety of student-scientist projects throughout the world. In addition to topics covered, these programs differ significantly in the extent of interaction among educators, professionals, and students. Please request further information from the program's education contact. Information on programs is accurate as of May 1997.

Aquanaut Program
Students are part of a marine science research team.

Education Contact: Ivar G. Babb
Avery Point Campus
NURC
1084 Shennecossett Road
Groton, CT 06340
860-405-9121
e-mail: nurcadm1@uconnvm.uconn.edu
homepage: http://www.ucc.uconn.edu/~wwwnurc

Athena
Students analyze Internet-available geophysical information.

Education Contact: Kathee Terry
10929 NE 119th Street
Kirkland, WA 98034
206-823-2997
e-mail: kterry@seanet.com
homepage: http://athena.wednet.edu

BirdWatch
Students examine bird populations and behaviors.

Education Contact: Elissa Landre
Broadmoor Wildlife Sanctuary
280 Eliot Street
South Natick, MA 01760
508-655-2296
e-mail: landre@whale.simmons.edu
homepage: http://earth.simmons.edu/birdwatch/birdwatch.html

Boreal Forest Watch
Students perform environmental monitoring on study plots in northern Saskatchewan.

Education Contact: Brad Muir
Prince Albert National Park
Box 100
Wakesiu Lake, Saskatchewan
Canada S0J 2Y0
306-663-4514
e-mail: brad_muir@pch.gc.ca
homepage: http://pyramid.sr.unh.edu/csrc/bfw/

Declining Amphibian Population Task Force (DAPTF)
Students monitor population fluctuations in native amphibians.

Education Contact: Tim Halliday
Department of Biology
The Open University
Milton Keynes MK7 6AA
United Kingdom
011 0 1908 653831
e-mail: t.r.halliday@open.ac.uk
homepage: http://www.open.ac.uk/info/newsletters/FROGLOG.html

Desert and Desertification Project
Students collect information on various features of deserts and the processes of desert formation.

Education Contact: Hannah Sivan
High School for Environmental Studies
Midreshet Ben Gurion
The Negev, Israel
e-mail: hannahs@environment.negev.k12.il
homepage: http://environment.negev.k12.il/desert/desert.htm

Earth System Science Community (ESSC)
Students use Internet-available data to research questions about the earth/space system.

Education Contact: Farzad Mahootian
ECOlogic Corporation
19 Eye Street, NW
Washington, DC 20001
202-218-4100
e-mail: farzad@twix.ecologic.net
homepage: http://www.circles.org

Earthwatch Education Program
Students participate in a variety of international field-based research projects.

Education Contact: Tally Forbes
680 Mt. Auburn Street
Watertown, MA 02272-9104
617-926-8200
e-mail: tforbes@earthwatch.org
homepage: http://atlas.earthwatch.org/ed/

EcoWatch
Students collect a variety of environmental data to examine local environmental quality.

Education Contact: Bill Clemens
8 Hunt Club Drive
Scarborough, Ontario
Canada M1N 2W8
416-690-8264
e-mail: bill_clemens@sbe.scarborough.on.ca
homepage: no website

FeederWatch
Students observe birds at feeders to examine population fluctuations.

Education Contact: Margaret Barker
Cornell Lab of Ornithology
159 Sapsucker Woods Road
Ithaca, NY 14850
607-254-2414
e-mail: nwd1@cornell.edu
homepage: http://www.ornith.cornell.edu

ForestWatch
Students observe the health of white pine trees as an indication of overall ecosystem health.

Education Contact: Barry Rock
Complex Systems Research Center
Morse Hall
University of New Hampshire
Durham, NH 03824
603-862-1792
e-mail: barry.rock@unh.edu
homepage: no website

Give Water a Hand
Students design water quality projects of local importance.

Education Contact: Kadi Row
216 Agriculture Hall
1450 Linden Drive
Madison, WI 53706
800-928-3720
e-mail: krow@facstaff.wisc.edu
homepage: http://www.uwex.edu/erc

GLOBE
Students collect local climate, hydrology, and ecosystem data for integration
into a global visualization.

Education Contact: Ralph Coppola
744 Jackson Place, NW
Washington, DC 20503
202-395-7600
e-mail: rcoppola@globe.gov
homepage: http://www.globe.gov

GREEN
Students examine local water quality and discuss policy implications.

206 South 5th Avenue, Suite 150
Ann Arbor, MI 48104
313-761-8142
e-mail: green@green.org
homepage: http://www.igc.apc.org/green

Hands On Universe
Students search for nearby and distant supernovas.

Education Contact: Carl Pennypacker
Lawrence Berkeley Laboratory
Building 50, Room 236
Berkeley, CA 94720
510-486-7429
e-mail: pennypacker@lbl.gov
homepage: http://hou.lbl.gov

High School Human Genome Project
Students determine human DNA sequences and discuss the ethical and social
implications of the Human Genome Project.

Education Contact: Maureen Munn
Department of Molecular Biotechnology
University of Washington
Box 352145
Seattle, WA 98195
206-616-4538
e-mail: mmunn@u.washington.edu
homepage: http://chroma.mbt.washington.edu

Hydrogen Isotope Study
Students collect samples of portions of a food chain to compare ratios of various hydrogen isotopes.

Education Contact: Orley Taylor
Department of Entomology
7005 Haworth Hall
University of Kansas
Lawrence, KS 66045
913-864-4051
e-mail: monarch@ukans.edu
homepage: http://www.keil.ukans.edu/~monarch/

Journey North
Students observe wildlife migration and seasonal changes.

Education Contact: Elizabeth Donnelly
18150 Breezy Point Road
Wayzata, MN 55391
612-476-6470
e-mail: jnorth@learner.org
homepage: http://www.learner.org/k12

Kids as Global Scientists
Students study local and regional weather actors.

Education Contact:

Nancy Songer and/or Holly DeVaul
University of Colorado–Boulder
Campus Box 249
Boulder, CO 80309-0249
303-492-3424
e-mail: kgs@spot.colorado.edu
homepage: http://www-kgs.colorado.edu

Mars Exploration
Students recreate the scientific process as it is being used in the exploration of
Mars.

Education Contact: Cheick Diarra
Education and Public Outreach
JPL
4800 Oak Grove Drive
Pasadena, CA 91109
818-354-6111
e-mail: cheick.m.diarra@jpl.nasa.gov
homepage: http://www.jpl.nasa.gov/mars/

Merrimack River Watershed Education Project
Students study a local river system.

Education Contact: Laura Ryder
2 Hazen Drive
Concord, NH 03301
603-271-3211
e-mail: no e-mail address available
homepage: http://www.wildlife.state.nh.us

Monarch Watch
Students follow monarch migrations from Canada to Mexico.

Education Contact: Orley Taylor
Department of Entomology
7005 Haworth Hall
University of Kansas
Lawrence, KS 66045
913-864-4051
e-mail: monarch@ukans.edu
homepage: http://www.keil.ukans.edu/~monarch/

Natural Resources, Science, Mathematics and Engineering Program
Students examine environmental factors at the Savannah River nuclear power plant site.

Education Contact: Norm Rischbieter
171 University Parkway, Box 3
Aiken, SC 29801
803-725-0070
e-mail: normr@aiken.sc.edu
homepage: no website

Particles and Prairies
Students document longitudinal changes in a reconstructed prairie.

Education Contact: David Abler
PO Box 500
Batavia, IL 60510
630-840-5059
e-mail: dabler@fnal.gov
homepage: http://www-ed.fnal.gov/

Project del Rio
US and Mexican students study water quality along the Rio Grande.

Education Contact: Lisa LaRocque
1494A South Solano
Los Cruces, NM 88001
505-522-7511
e-mail: llarocque@igc.org
homepage: http://www.econet.apc.org/green/delrio.html

Project PigeonWatch
Students observe pigeon colors and behavior in city flocks.

Education Contact: Martha Fischer
Cornell Lab of Ornithology
159 Sapsucker Woods Road
Ithaca, NY 14850
607-254-2403
e-mail: pigeonwatch@cornell.edu
homepage: http://www.ornith.cornell.edu/cs/ppw/main.html

Project Tanager
Students observe the fluctuations in populations of birds in forested landscapes.

Education Contact: Margaret Barker
Cornell Lab of Ornithology
159 Sapsucker Woods Road
Ithaca, NY 14850
607-254-2440
e-mail: nwd1@cornell.edu
homepage: http://www.ornith.cornell.edu

Rivers Curriculum Project
Students monitor the chemistry, biology and geology of local rivers.

Education Contact: Robert Williams
Southern Illinois University at Edwardsville
Box 2222
Edwardsville, IL 62026-2222
618-692-3788
e-mail: rivers@siue.edu
homepage: http://www.siue.edu/OSME/river/river.html

Rivers of Colorado Water Watch Network
Students evaluate local watershed and stream ecology.

Education Contact: Barb Horn
6060 Broadway
Denver, CO 80216
303-291-7388
e-mail: no e-mail address available
homepage: no website

Saltwatch
Students determine salinity of surface and ground water samples in Victoria, Australia.

Education Contact: Tanya Kruger
e-mail: tkruger@peg.apc.org
homepage: http://www.dse.vic.gov.au /saltwatc.htm

SITE (Students Investigating Today's Environment)
Students analyze water from the Snake River and its tributaries.

Education Contact: Robert Beckwith
Eagle High School
574 Park Lane
Eagle, ID 83616-4512
208-939-2189 ext. 303
e-mail: reb@site.eitc.edu
homepage: no website

SNAP (School Nature Area Project)
Students collect snow depth data.

Education Contact: Bill Lindquist
1520 St. Olaf Avenue
Northfield, MN 55057
507-646-3977
e-mail: lindquis@stolaf.edu
homepage: http://www.stolaf.edu/other/snap/

SPAN (Sun Photometer Atmospheric Network)
Students measure local ozone variations.

Education Contact: Forrest M. Mims III
433 Twin Oak Road
Seguin, TX 78155
210-372-0548
e-mail: fmims@aol.com
homepage: http://www.concord.org/haze/

SPaRCE (Schools of the Pacific Rainfall Climate Experiment)
Students participate in global climate research by measuring rainfall.

Education Contact: Mark Morrissey
University of Oklahoma
100 East Boyd, #1210
Norman, OK 73019
405-325-3207
e-mail: sparce@hoth.gcn.ou.edu
homepage: http://radar.metr.uoknor.edu/sparce/sparce.html

Stream Team
Students survey water quality in local streams.

Education Contact: Sharon Meigs
Prince George's County Programs and Planning Division
9400 Peppercorn Place, Suite 540
Englewood Center Three
Largo, MD 20774
301-883-5898
e-mail: solomon@gcmd.gsfc.nasa.gov
homepage: no website

Student Research Apprenticeship Program
Students are members of scientists' research teams.

Education Contact: Jeff Estes
Pacific Northwest Lab-OUSEP
PO Box 999
Mail Stop K1-22
Richland, WA 99352
509-375-2820
e-mail: jc_estes@pnl.gov
homepage: http://www.pnl.gov:2080/education/index.html

Student Watershed Research Program
Students collect physical, biological, and chemical data on local watersheds.

Education Contact: Jane Blair
Saturday Academy—OGI
PO Box 91000
Portland, OR 97291-1000
503-690-1344
e-mail: satacad@admin.ogi.edu
homepage: http://www.ogi.edu:80/ satacad

SUN (Students Understanding Nature)
Students measure solar energy, especially ultraviolet-B.

Education Contact: Gil Yanow
4800 Oak Grove Drive
Mail Stop CS-530
Pasadena, CA 91109
818-354-6916
e-mail: gilbert.yanow@jpl.nasa.gov
homepage: http://sunshine.jpl.nasa.gov/

WhaleNet
Students study whale behavior.

Education Contact: J. Michael Williamson
Wheelock College - 200 The Riverway
Boston, MA 02215-4176
617-734-5200 ext. 256
e-mail: williams@whale.wheelock.edu
homepage: http://whale.wheelock.edu/

Wildlife Watch
Students take part in national environmental data-collection projects.

Education Contact: Mary Cornwell
The Green, Witham Park
Waterside South, Lincoln LN5 7JR
United Kingdom
01522-544400
e-mail: no e-mail address available
homepage: http://www.youthnet.org.uk

Winter Weather and Whitetail Deer
Student-collected data on snow depth is used by the NH Fish and Game Department to develop wildlife management strategies.

Education Contact: Judy Silverberg
2 Hazen Drive
Concord, NH 03301
603-271-3211
e-mail: jsilverb@wildlife.state.nh.us
homepage: no website

Index